Waste Management Handbook

Waste Management Handbook

Lydia Hemsworth

Callisto Reference

www.callistoreference.com

Callisto Reference,
118-35 Queens Blvd., Suite 400,
Forest Hills, NY 11375, USA

Visit us on the World Wide Web at:
www.callistoreference.com

ISBN: 978-1-64116-626-3 (Hardback)

Cataloging-in-Publication Data

Waste management handbook / Lydia Hemsworth.
p. cm.
Includes bibliographical references and index.
ISBN 978-1-64116-626-3
1. Refuse and refuse disposal. 2. Sanitation. 3. Factory and trade waste.
4. Salvage (Waste, etc.). I. Hemsworth, Lydia.

TD793 .W37 2022
363.728--dc23

Table of Contents

Preface VII

Chapter 1 Waste and its Types 1

- Biodegradable Waste 2
- Sewage 3
- Green Waste 7
- Food Waste 9
- Sewage Sludge 17
- Electronic Waste or E-Waste 23
- Industrial Waste 41
- Radioactive Waste 42

Chapter 2 Types of Waste Management 60

- Waste Management 60
- Solid Waste Management 64
- Liquid Waste Management 79
- Agricultural Waste Management 88
- Organic Waste Management 121
- Slaughterhouse Waste Management 124
- Hazardous Waste Management 136
- High-level Radioactive Waste Management 147

Chapter 3 Principles of Waste Management 152

- Waste Minimization 154
- Waste Hierarchy 159
- Precycling 162
- Zero Waste 167
- Resource Efficiency 177
- Polluter Pays Principle 178
- Extended Producer Responsibility 180
- Pay as you Throw 186
- Energy Recovery from Waste 187

Chapter 4 Waste Handling and Disposal 189
- Waste Sorting 190
- Kerbside Collection 196
- Waste Container 204
- Disposal Methods 205

Chapter 5 Recycling and Re-use of Waste 210
- Waste Recycling 210
- Recyclates 226
- Plastic Recycling 227
- Reuse 231
- Waste Exchange 234
- Reusable Packaging 237
- Remanufacturing 239
- Recycling of Industrial Waste 242

Permissions

Index

Preface

The purpose of this book is to help students understand the fundamental concepts of this discipline. It is designed to motivate students to learn and prosper. I am grateful for the support of my colleagues. I would also like to acknowledge the encouragement of my family.

Waste management refers to the actions and activities through which waste is managed from its origin to its final disposal. It involves the collection, transportation, treatment and disposal of waste. Different kinds of wastes such as solid, liquid or gaseous waste are treated and disposed through different management processes. Industries and households are the major sources of chemical and biological wastes. Waste management aims to reduce the harmful effects of waste, and protect human health and the environment. Landfill and incineration are two major disposal methods used for waste management. It also includes recycling and re-using waste. Recycling refers to the collection and re-use of waste materials. Re-use includes various processes such as biological reprocessing, energy recovery and pyrolysis. This book provides comprehensive insights into the field of waste management. Most of the topics introduced herein cover new techniques of waste management. This book will provide comprehensive knowledge to the readers.

A foreword for all the chapters is provided below:

Chapter - Waste and its Types

The substance which has been discarded after use, or is worthless, defective and of no use, is generally termed as waste. The different kinds of wastes include biodegradable waste, electronic waste, sewage sludge, radioactive waste, green waste, etc. This is an introductory chapter which will introduce briefly all the significant types of waste.

Chapter - Types of Waste Management

Waste management includes the activities and methods used to manage different types of waste. Solid waste management, agricultural waste management, slaughterhouse waste management and organic waste management are the various types of waste management. The chapter closely examines these types of waste management to provide an extensive understanding of the subject.

Chapter - Principles of Waste Management

The fundamental principles of waste management include waste minimization, resource efficiency, polluter pays principle, waste hierarchy, precycling, energy recovery and extended producer responsibility. All these diverse principles of waste management have been carefully analyzed in this chapter.

Chapter - Waste Handling and Disposal

Some of the different processes which are a part of waste handling and disposal are waste sorting, kerbside collection, incineration and waste compaction. This chapter closely examines these methods of waste handling and disposal to provide an extensive understanding of the subject.

Chapter - Recycling and Re-use of Waste

The process of converting waste material into new objects and materials is known as recycling. It helps in the reduction of greenhouse gas emissions. Reusing waste materials involves the use of waste product without further transformation and without changing its shape or original nature. This chapter discusses in detail the concepts related to waste recycling and reuse such as waste exchange and remanufacturing.

Lydia Hemsworth

1

Waste and its Types

The substance which has been discarded after use, or is worthless, defective and of no use, is generally termed as waste. The different kinds of wastes include biodegradable waste, electronic waste, sewage sludge, radioactive waste, green waste, etc. This is an introductory chapter which will introduce briefly all the significant types of waste.

A natural part of the life cycle, waste occurs when any organism returns substances to the environment. Living things take in raw materials and excrete wastes that are recycled by other living organisms. However, humans produce an additional flow of material residues that would overload the capacity of natural recycling processes, so these wastes must be managed in order to reduce their effect on our aesthetics, health, or the environment.

Solid and fluid, hazardous and non-toxic wastes are generated in our households, offices, schools, hospitals, and industries. No society is immune from day-to-day issues associated with waste disposal. How waste is handled often depends on its source and characteristics, as well as any local, state, and federal regulations that govern its management. Practices generally differ for residences and industries, in urban and rural areas, and for developed and developing countries.

Municipal Solid Waste

Waste collected from residences, commercial buildings, institutions such as hospitals and schools, and light industrial operations is most often categorized as municipal solid waste. MSW consists primarily of paper, containers and packaging, food wastes, yard trimmings, and other inorganic wastes. Municipal solid waste can also include industrial sludge, classified as hazardous or non-hazardous, resulting from a wide array of mining, construction, and manufacturing processes.

In 2006, Americans generated more than 250 million tons of trash. Nearly 33 percent, 82 million tons of materials, was recycled; the energy equivalent of more than 10 billion gallons of gasoline. More than 30 million tons (12.5 percent) were combusted through an energy recovery process, and approximately 138 million tons (55 percent) of materials were discarded in landfills. Municipal waste, when properly managed, does not pose an immediate threat to human health or the environment.

Hazardous Waste

Waste material that is flammable, corrosive, reactive, or toxic, which can be in the form of a solid, liquid, or gas—is defined as hazardous waste. Although the term often evokes an image of items marked with skull and crossbones, many hazardous wastes include products used every day, including paint, used oil from cars, batteries, shoe polish, and even laundry detergent. In addition, many of the items that we rely upon generate hazardous waste during the process of their production.

The U.S. Environmental Protection Agency (EPA) reported that 279 million tons of hazardous wastes were generated in 1996; ninety-six percent of which was industrial process water by-product waste. In 1997 the EPA made a rule change that separated industrial wastewater from hazardous waste reporting. This change can clearly be seen in current reporting numbers; in 2005 the EPA reported hazardous waste generation of just under 38.5 million tons.

Businesses that generate hazardous wastes are required by legislation to manage them from generation to disposal. The waste is often treated to change its biological, chemical, or physical characteristics in order to make it less hazardous or to reduce its overall volume. Some hazardous materials can be recycled if it is environmentally safe to do so, although it can be expensive. Any leftover waste is then safely disposed of to further neutralize any adverse affects to human health or the environment. Today, many industries are attempting to reduce their generation of hazardous waste by modifying their manufacturing processes or by replacing hazardous materials with less hazardous or non-hazardous substitutes.

E-waste

A new and growing segment of our waste stream is termed 'e-waste.' Although not clearly defined, e-waste applies to much of the electronic equipment used by businesses and individual consumers that are nearing the end of their usefulness. This includes, but is not limited to, computers, fax machines, copiers, and televisions.

A factor that complicates disposal of these items is that certain components contain hazardous materials. The cathode ray tubes in computer monitors and televisions are an example. As such, many old electronics sit idle due to the uncertainty of how to manage them.

Yet, many of these products can be reused, refurbished, or recycled. There are a number of options to reuse equipment that can still function—from selling it to someone who can use it or giving it as a charitable donation. Many non-working items can be refurbished in order to recover them into working condition. Finally, any components that cannot be repaired can often be recycled.

Although there no federal regulation for e-waste, some states have taken various approaches toward its management. Many states, including Massachusetts, Florida, and New York, have streamlined their regulations to increase the level of recycling. On the other hand, in 1993 California passed the Electronic Waste Recycling Act; yet, they also regulate cathode ray tubes as hazardous waste, banning them from regular trash disposal.

BIODEGRADABLE WASTE

Biodegradable waste is a type of waste, typically originating from plant or animal sources, which may be degraded by other living organisms. Waste that cannot be broken down by other living organisms are called non-biodegradable.

Biodegradable waste can be commonly found in municipal solid waste (sometimes called biodegradable municipal waste, or BMW) as green waste, food waste, paper waste, and biodegradable plastics. Other biodegradable wastes include human waste, manure, sewage, slaughterhouse waste. In the absence of oxygen much of this waste will decay to methane by anaerobic digestion.

A container for collecting the biodegradable waste of a household.

Climate Change Impacts

The main environmental threat from biodegradable waste is the production of methane in landfills. Methane is 21 times as potent a greenhouse gas as carbon dioxide and accounted for some 3% of total greenhouse gas emissions in the EU-15 in 1995. The Landfill Directive 1999/31/EC obliges Member States to reduce the amount of biodegradable waste that they landfill to 35% of 1995 levels by 2016, which will significantly reduce the problem.

Uses of Biodegradable Waste

Biodegradable waste can often be used for composting which turns the waste into humus. It can also be used as a resource for heat, electricity and fuel in future, by using anaerobic digestion as it is being achieved by the swiss Kompogas treatment for 20 years now. This produces additional Biogas and still delivers usable nutrients that can be implemented to the soil. Especially animal-based kitchen scraps (such as leftover fish, meat), are best disposed this way (as these are best not composted). Finally note that leftover human food (vegetables, no animal flesh) can also be used as fodder to flightless birds, goats, pigs, sheep, or alternatively can be fed to birds, and other wildlife.

In the USA, approximately 50% of waste sent to landfills or incinerators is biodegradable; therefore, composting could create many new jobs, and recover significant resources that are being lost.

SEWAGE

Sewage (or domestic wastewater or municipal wastewater) is a type of wastewater that is produced by a community of people. It is characterized by volume or rate of flow, physical condition, chemical and toxic constituents, and its bacteriologic status (which organisms it contains and in what quantities). It consists mostly of greywater (from sinks, tubs, showers, dishwashers, and clothes washers), blackwater (the water used to flush toilets, combined with the human waste that it flushes away); soaps and detergents; and toilet paper (less so in regions where bidets are widely used instead of paper).

Sewage usually travels from a building's plumbing either into a sewer, which will carry it elsewhere, or into an onsite sewage facility (of which there are many kinds). Whether it is combined with surface runoff in the sewer depends on the sewer design (sanitary sewer or combined sewer). The reality is, however, that most wastewater produced globally remains untreated causing widespread water pollution, especially in low-income countries: A global estimate by UNDP and UN-Habitat is that 90% of all wastewater generated is released into the environment untreated. In many developing countries the bulk of domestic and industrial wastewater is discharged without any treatment or after primary treatment only.

Raw sewage arriving at a sewage treatment plant.

The term sewage is nowadays regarded as an older term and is being more and more replaced by "wastewater". In general American English usage, the terms "sewage" and "sewerage" mean the same thing. In common British usage, and in American technical and professional English usage, "sewerage" refers to the infrastructure that conveys sewage.

Types

Greywater samples before and after treatment in a constructed wetland.

The wastewater from residences and institutions, carrying bodily wastes (primarily feces and urine), washing water, food preparation wastes, laundry wastes, and other waste products of normal living, are classed as domestic or sanitary sewage.

Liquid-carried wastes from stores and service establishments serving the immediate community, termed commercial wastes, are included in the sanitary or domestic sewage category if their characteristics are similar to household flows. Wastes that result from industrial processes such as the production or manufacture of goods are classed as industrial wastewater, not as sewage.

Surface runoff, also known as storm flow or overland flow, is that portion of precipitation that runs rapidly over the ground surface to a defined channel. Precipitation absorbs gases and particulates from the atmosphere, dissolves and leaches materials from vegetation and soil, suspends matter from the land, washes spills and debris from urban streets and highways, and carries all these pollutants as wastes in its flow to a collection point.

Pollutants

Organic Pollutants and Nutrients

Sewage is a complex mixture of chemicals, with many distinctive chemical characteristics. These include high concentrations of ammonium, nitrate, nitrogen, phosphorus, high conductivity (due to high dissolved solids), high alkalinity, with pH typically ranging between 7 and 8. The organic matter of sewage is measured by determining its biological oxygen demand (BOD) or the chemical oxygen demand (COD).

Pathogens

Sewage contains human feces, and therefore often contains pathogens of one of the four types:

- Bacteria (for example Salmonella, Shigella, Campylobacter, Vibrio cholerae),
- Viruses (for example hepatitis A, rotavirus, enteroviruses),
- Protozoa (for example Entamoeba histolytica, Giardia lamblia, Cryptosporidium parvum),
- Parasites such as helminths and their eggs (e.g. ascaris (roundworm), ancylostoma (hookworm), trichuris (whipworm)).

Sewage can be monitored for both disease-causing and benign organisms with a variety of techniques. Traditional techniques involve filtering, staining, and examining samples under a microscope. Much more sensitive and specific testing can be accomplished with DNA sequencing, such as when looking for rare organisms, attempting eradication, testing specifically for drug-resistant strains, or discovering new species. Sequencing DNA from an environmental sample is known as metagenomics.

Micro-pollutants

Sewage also contains environmental persistent pharmaceutical pollutants. Trihalomethanes can also be present as a result of past disinfection.

Sewage has also been analyzed to determine relative rates of use of prescription and illegal drugs among municipal populations.

Health and Environmental Aspects

All categories of sewage are likely to carry pathogenic organisms that can transmit disease to humans and animals. Sewage also contains organic matter that can cause odor and attract flies. Sewage contains nutrients that may cause eutrophication of receiving water bodies; and can lead to ecotoxicity.

Collection

A medieval waste pipe in Stockholm Old Town formerly deposited sewage on the street to be flushed away by rain.

Sewage canal of a medieval house as depicted in 1447 St. Barbara Altarpiece in the National Museum in Warsaw.

A system of sewer pipes (sewers) collects sewage and takes it for treatment or disposal. The system of sewers is called sewerage or sewerage system in British English and sewage system in American English. Where a main sewerage system has not been provided, sewage may be collected from homes by pipes into septic tanks or cesspits, where it may be treated or collected in vehicles and taken for treatment or disposal. Properly functioning septic tanks require emptying every 2–5 years depending on the load of the system.

Treatment

Sewage treatment is the process of removing the contaminants from sewage to produce liquid and solid (sludge) suitable for discharge to the environment or for reuse. It is a form of waste management. A septic tank or other on-site wastewater treatment system such as biofilters or constructed wetlands can be used to treat sewage close to where it is created.

Sewage treatment results in sewage sludge which requires sewage sludge treatment before safe disposal or reuse. Under certain circumstances, the treated sewage sludge might be termed "biosolids" and can be used as a fertilizer.

In developed countries sewage collection and treatment is typically subject to local and national regulations and standards.

Disposal

Raw sewage is also disposed of to rivers, streams, and the sea in many parts of the world. Doing so can lead to serious pollution of the receiving water. This is common in developing countries and may still occur in some developed countries, for various reasons – usually related to costs.

Ships at sea are forbidden from discharging their sewage overboard unless three miles or more from shore.

Reuse of Treated or Untreated Sewage

Increasingly, agriculture is using untreated wastewater for irrigation. Cities provide lucrative markets for fresh produce, so are attractive to farmers. However, because agriculture has to compete for increasingly scarce water resources with industry and municipal users, there is often no alternative for farmers but to use water polluted with urban waste, including sewage, directly to water their crops. There can be significant health hazards related to using water loaded with pathogens in this way, especially if people eat raw vegetables that have been irrigated with the polluted water.

The International Water Management Institute has worked in India, Pakistan, Vietnam, Ghana, Ethiopia, Mexico and other countries on various projects aimed at assessing and reducing risks of wastewater irrigation. They advocate a 'multiple-barrier' approach to wastewater use, where farmers are encouraged to adopt various risk-reducing behaviours. These include ceasing irrigation a few days before harvesting to allow pathogens to die off in the sunlight, applying water carefully so it does not contaminate leaves likely to be eaten raw, cleaning vegetables with disinfectant or allowing fecal sludge used in farming to dry before being used as a human manure. The World Health Organization has developed guidelines for safe water use.

GREEN WASTE

Green waste, also known as "biological waste," is any organic waste that can be composted. It is most usually composed of refuse from gardens such as grass clippings or leaves, and domestic or industrial kitchen wastes. Green waste does not include things such dried leaves, pine straw, or hay. Such materials are rich in carbon and considered "brown wastes," while green wastes contain high in concentrations of nitrogen. Green waste can be used to increase the efficiency of many composting operations and can be added to soil to sustain local nutrient cycling.

Collection of Green Waste

Green waste can be collected via municipal curbside collection schemes or through private waste management businesses. Many communities, especially in the United Kingdom, have initiated green waste recycling and collection programs in order to decrease the amount of biodegradable materials in landfills. Communities are provided with, or can provide their own, compost receptacles that they fill with plant and food remains, which are then emptied on a regular basis. Programs such as this allow communities to be an active part of composting their green waste which allows them to play an active role in decreasing the amount of food being dumped into local and regional landfills.

This is a bin filled with materials that comprise green waste, such as kitchen wastes and plant trimmings.

Uses

Green waste can be used to improve the quality and sustainability of industrially manufactured topsoils and the sanitariness and efficiency of sewage disposal.

Manufactured Topsoils

Green waste is an integral part of many manufactured topsoils, as it provides both nutrients for growing plants and increases the volume of manufactured topsoils. Its woody components do not decompose quickly, so they provide the bulk that is necessary for supplementary topsoils. Mixing industrial wastes such as fly ash or coal dust with green waste to create artificial topsoil not only facilitates the repurposing of industrial debris and keeps it out of landfills, but it also allows the nutrients in green waste materials to be cycled back into the environment. By utilizing fly ash in conjunction with green waste, manufactured soils are able to increase their water holding capacity while simultaneously recycling refuse materials that might otherwise take up space in a landfill. This allows fly ash and green waste to increase local soil nutrient levels and promotes natural nutrient cycling processes.

Sewage Disposal

Green waste can also be mixed with sewage wastes and composted, providing a safe, environmentally sustainable option for sewage disposal. Co-composting green and sewage wastes eliminates the risk that pathogens and pollutants contained in sewage wastes might pose to the environment. Utilization of green waste to dispose of sewage wastes not only decreases the amount of sewage incinerated and dumped each year, it also facilitates the cycling of organic wastes back into the environment. The co-composted remains of these organic wastes can be safely employed for use in agriculture. This process decreases the amount of trash being dumped into landfills and other trash repositories and allows for the complete cycling of organic nutrients through the environment.

Renewable Energy

Vegetable waste being dumped in a market.

Biogas captured from biodegradable green waste can be used as biofuel. Green waste can be composed of non-food crops, which decompose to produce cellulosic ethanol. It can also help reduce the necessity of petroleum gases, which produce large amounts of greenhouse gases, such as carbon dioxide, when burned.

Soil Health

Green waste composting has also been linked to suppression of soil borne diseases such as damping off and root rots that affect large agricultural ventures like greenhouses and large-scale farms. This disease suppressive quality has positive implications for lesser-developed nations that do not have the technology or resources to purchase expensive fertilizers.

Addition of composts which contain residues and particles of heavy metals to soil can raise the soil's heavy metal content and increase the probability of these metals to transfer into crop plants. When biological, or green waste is added to these soil samples, plant uptake of heavy metal has been shown to decrease crop uptake of metals compared to other types of compost composed of things such as sewage sludge. This can protect consumers and the environment from biomagnification caused by long-term accumulation of heavy metal particles within the soil and plant life of an area.

FOOD WASTE

Food waste or food loss is food that is discarded or lost uneaten. The causes of food waste or loss are numerous and occur at the stages of producing, processing, retailing and consuming.

Global food loss and waste amount to between one-third and one-half of all food produced. Loss and wastage occur at all stages of the food supply chain or value chain. In low-income countries, most loss occurs during production, while in developed countries much food – about 100 kilograms (220 lb) per person per year – is wasted at the consumption stage.

Fruit and vegetables in a dumpster, discarded uneaten.

Causes

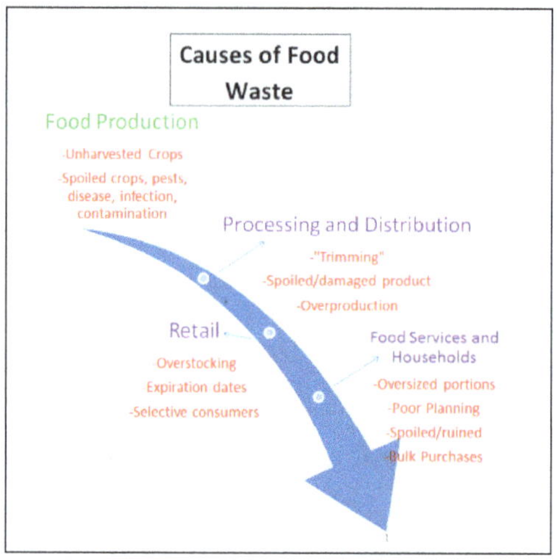

Production

In the US, food waste can occur at most stages of the food industry and in significant amounts. In subsistence agriculture, the amounts of food waste are unknown, but are likely to be insignificant by comparison, due to the limited stages at which waste can occur, and given that food is grown for projected need as opposed to a global marketplace demand. Nevertheless, on-farm losses in storage in developing countries, particularly in African countries, can be high although the exact nature of such losses is much debated.

In the food industry of the United States, the food supply of which is the most diverse and abundant of any country in the world, waste occurs from the beginning of food production chain. From planting, crops can be subjected to pest infestations and severe weather, which cause losses before harvest. Since natural forces (e.g. temperature and precipitation) remain the primary drivers of crop growth, losses from these can be experienced by all forms of outdoor agriculture. On average, farms in the United States lose up to six billion pounds of crops every year because of these unpredictable conditions. The use of machinery in harvesting can cause waste, as harvesters may be unable to discern between ripe and immature crops, or collect only part of a crop. Economic factors, such as regulations and standards for quality and appearance, also cause food waste; farmers often harvest selectively, preferring to leave crops not to standard in the field (where they can be used as fertilizer or animal feed), since they would otherwise be discarded later. This method of removing undesirable produce from harvest collection, distribution sites and grocery stores is called Culling. The USDA defines culling as "the individual removal of genetically undesirable, inferior, weak, diseased, or infested plants from a planting in order to ensure the level of genetic purity or vigor of the crop". However, usually when culling occurs at the production, food processing, retail and consumption stages, it is to remove or dispose of produce with a strange or imperfect appearance rather than produce that is spoiled or unsafe to eat. In urban areas, fruit and nut trees often go unharvested because people either don't realize that the fruit is edible or they fear that it is contaminated, despite research which shows that urban fruit is safe to consume.

Food Processing

Food waste continues in the post-harvest stage, but the amounts of post-harvest loss involved are relatively unknown and difficult to estimate. Regardless, the variety of factors that contribute to food waste, both biological/environmental and socio-economical, would limit the usefulness and reliability of general figures. In storage, considerable quantitative losses can be attributed to pests and micro-organisms. This is a particular problem for countries that experience a combination of heat (around 30 °C) and ambient humidity (between 70 and 90 per cent), as such conditions encourage the reproduction of insect pests and micro-organisms. Losses in the nutritional value, caloric value and edibility of crops, by extremes of temperature, humidity or the action of micro-organisms, also account for food waste. Further losses are generated in the handling of food and by shrinkage in weight or volume.

Some of the food waste produced by processing can be difficult to reduce without affecting the quality of the finished product. Food safety regulations are able to claim foods which contradict standards before they reach markets. Although this can conflict with efforts to reuse food waste (such as in animal feed), safety regulations are in place to ensure the health of the consumer; they are vitally important, especially in the processing of foodstuffs of animal origin (e.g. meat and dairy products), as contaminated products from these sources can lead to and are associated with microbiological and chemical hazards.

Retail

Packaging protects food from damage during its transportation from farms and factories via warehouses to retailing, as well as preserving its freshness upon arrival. Although it avoids considerable food waste, packaging can compromise efforts to reduce food waste in other ways, such as by contaminating waste that could be used for animal feedstocks.

In 2013 the non-profit Natural Resources Defense Council (NRDC) performed research that they state suggests that the leading cause of food waste in America is due to uncertainty over food expiration dates, such as confusion in deciphering best before, sell-by or use-by dates. Joined by Harvard Law, the NRDC produced a study called The Dating Game: How Confusing Food Date Labels Leads to Food Waste in America. This United States-based study looked at the intertwining laws which lead labeling to end up unclear and erratic. This uncertainty leads to consumers to toss food, most often because they think the food may be unsafe or misunderstand the labeling on the food completely. Lack of regulation on labeling can result in large quantities of food being removed from the market overall.

Discarded bagels.

Retail stores throw away large quantities of food. Usually, this consists of items that have reached either their best before, sell-by or use-by dates. Food that has passed the best before, and sell-by date, and even some food that passed the use-by date is still edible at the time of disposal, but stores have widely varying policies to handle the excess food. Some stores put effort into preventing access to poor or homeless people, while others work with charitable organizations to distribute food. Retailers also contribute to waste as a result of their contractual arrangements with suppliers. Failure to supply agreed quantities renders farmers or processors liable to have their contracts cancelled. As a consequence, they plan to produce more than actually required to meet the contract, to have a margin of error. Surplus production is often simply disposed of.

Retailers usually have strict cosmetic standards for produce, and if fruits or vegetables are misshapen or superficially bruised, they are often not put on the shelf. In the United States, an estimated six billion pounds of produce is wasted each year because of its appearance. The USDA publishes guidelines used as a baseline assessment by produce distributors, grocery stores, restaurants and other consumers in order to rate the quality of food. These guidelines and how they rate are readily available on their website. For example, apples get graded by their size, color, wax residue, firmness, and skin appearance. If an apples rank highly in these categories and show close to no superficial defects, they are rated as "U.S. Extra Fancy" or "U.S. Fancy", these are the typical ratings sought out by grocery stores when purchasing their produce. Any apples with suboptimal levels of appearance are ranked as either "U.S. Number 1" or "Utility" and are not normally purchased for retail, as recommended by produce marketing sources, despite being safe and edible.

The fish industry also contributes to the annual amount of food waste because of cosmetic standards that the fish are held up to. Nearly "2.3 million tonnes of fishes are discarded in the North Atlantic and the North Sea each year." Approximately 40 to 60 percent of "all fish caught in Europe is discarded – either because they are the wrong size or species." Addressing this, there are many campaigns focused on raising retailer and consumer awareness about food that fails to meet certain standards for appearance.

Consumption

Consumers are directly and indirectly responsible for wasting a lot of food, which could for a large part be avoided if they were willing to accept suboptimal food (SOF) that deviates in sensory characteristics (odd shapes, discolourations) or has a best-before date that is approaching or has passed, but is still perfectly fine to eat. COSUS (COnsumers in a SUStainable food supply chain) is a SUSFOOD ERA-net research project under the topic 'Understanding consumer behaviour to encourage a more sustainable food choice'.

Impact on the Environment

According to the Food and Agriculture Organization (FAO), food waste is responsible for 8 percent of global man-made greenhouse gas emissions. The FAO concludes that nearly 30 percent of all available agricultural land in the world - 1.4 billion hectares - is used for produced but uneaten food. The global blue water footprint of food waste is 250 km3, that is the amount of water that flows annually through the Volga or 3 times Lake Geneva.

Reduction

Response to the problem of food waste at all social levels has varied hugely, including campaigns from advisory and environmental groups, and concentrated media attention on the subject.

Consumer Marketing

One way of dealing with food waste is to reduce its creation. Consumers can reduce spoilage by planning their food shopping, avoiding potentially wasteful spontaneous purchases, and storing foods properly. Widespread educational campaigns have been shown to be an effective way to reduce food waste. A British campaign called "Love Food Hate Waste" has raised awareness about preventative measures to address food waste for consumers. Through advertisements, information on food storage and preparation and in-store education, the UK observed a 21% decrease in avoidable household food waste over the course of 5 years. Another potential solution is for "smart packaging" which would indicate when food is spoiled more precisely than expiration dates currently do, for example with temperature-sensitive ink, plastic that changes color when exposed to oxygen, or gels that change color with time.

An initiative in Curitiba, Brazil called Cambio Verde allows farmers to provide surplus produce (produce they would otherwise discard due to too low prices) to people that bring glass and metal to recycling facilities (to encourage further waste reduction). In Europe, the Food Surplus Entrepreneurs Network (FSE Network), coordinates a network of social businesses and nonprofit initiatives with the goal to spread best practices to increase the use of surplus food and reduction of food waste.

Collection

In areas where the waste collection is a public function, food waste is usually managed by the same governmental organization as other waste collection. Most food waste is combined with general waste at the source. Separate collections, also known as source-separated organics, have the advantage that food wastes can be disposed of in ways not applicable to other wastes. In the United States, companies find higher and better uses for large commercial generators of food and beverage waste.

From the end of the 19th century through the middle of the 20th century, many municipalities collected food waste (called "garbage" as opposed to "trash") separately. This was typically disinfected by steaming and fed to pigs, either on private farms or in municipal piggeries.

Separate curbside collection of food waste is now being revived in some areas. To keep collection costs down and raise the rate of food waste segregation, some local authorities, especially in Europe, have introduced “alternate weekly collections” of biodegradable waste (including, e.g., garden waste), which enable a wider range of recyclable materials to be collected at reasonable cost, and improve their collection rates. However, they result in a two-week wait before the waste will be collected. The criticism is that particularly during hot weather, food waste rots and stinks, and attracts vermin. Waste container design is therefore essential to making such operations feasible. Curbside collection of food waste is also done in the U.S., some ways by combining food scraps and yard waste together. Several states in the U.S. have introduced a yard waste ban, not accepting leaves, brush, trimmings, etc. in landfills. Collection of food scraps and yard waste combined is then recycled and composted for reuse.

Disposal

As alternatives to landfill, food waste can be composted to produce soil and fertilizer, fed to animals, or used to produce energy or fuel.

Landfills and Greenhouse Gases

Dumping food waste in a landfill causes odour as it decomposes, attracts flies and vermin, and has the potential to add biological oxygen demand (BOD) to the leachate. The European Union Landfill Directive and Waste Regulations, like regulations in other countries, enjoin diverting organic wastes away from landfill disposal for these reasons. Starting in 2015, organic waste from New York City restaurants will be banned from landfills.

In countries such as the United States and the United Kingdom, food scraps constitute around 19% of the waste dumped in landfills, where it ends up rotting and producing methane, a greenhouse gas.

Methane, or CH4, is the second most prevalent greenhouse gas that is released into the air, also produced by landfills in the U.S. Although methane spends less time in the atmosphere (12 years) than CO2, it's more efficient at trapping radiation. It is 25 times greater to impact climate change than CO2 in a 100-year period. Humans accounts over 60% of methane emissions globally.

Animal Feed

Large quantities of fish, meat, dairy and grain are discarded at a global scale annually, when they can be used for things other than human consumption. The feeding of food scraps to domesticated animals is, historically, the most common way of dealing with household food waste. The animals turn roughly two thirds of their ingested food into gas or fecal waste, while the last third is digested and repurposed as meat or dairy products. There are also different ways of growing produce and feeding livestock that could ultimately reduce waste.

Bread and other cereal products discarded from the human food chain could be used to feed chickens. Chickens have traditionally been given mixtures of waste grains and milling by-products in a mixture called chicken scratch. As well, giving table scraps to backyard chickens is a large part of that movement's claim to sustainability, though not all backyard chicken growers recommend it. Ruminants and pigs have also been fed bakery waste for a long time.

Certain food waste (such as flesh) can also be used as feed in maggot farming. The maggots can then be fed to other animals.

Composting

Inevitable waste: peels of potato, onion, lemon, tangerine, banana, kiwi, egg.

Food waste can be biodegraded by composting, and reused to fertilize soil. Composting is the aerobic process completed by microorganisms in which the bacteria break down the food waste into simpler organic materials that can then be used in soil. By redistributing nutrients and high microbial populations, compost reduces water runoff and soil erosion by enhancing rainfall penetration, which has been shown to reduce the loss of sediment, nutrients, and pesticide losses to streams by 75–95%.

Traditional composting uses microbes to perform the decomposition, a process that is most efficient using low, mid, and high temperature microbes (psychrophilic, mesophilic, and thermophilic – respectively). The high temperatures required by the thermophilic microorganisms are hot enough to kill pathogens, making the product of this traditional composting satisfactory for use in soil according to the United States Environmental Protection Agency's (EPA) standards. The traditional decomposition process requires a long length of time and additional energy expended to turn the material frequently and maintain the aerobic process. Composting by thermophilic microbes can lead to nutrient loss and the compost product is heterogeneous, with the potential for higher levels of contaminants which can be harmful if used in agriculture. An alternate method of composting is vermicomposting.

Vermicomposting is the practise of feeding scraps to worms who produce fertilized soil as a byproduct. The process of composting using earth worms is completed in a short duration of time and requires no additional energy to turn and maintain aerobic processes, as these actions are already performed by the worms. In order to keep the worms alive, the environment has to be kept below 35° Celsius, therein making this compost ineligible for use in agriculture according to the EPA standards. The product of vermicomposting is homogenous and generally contains lower levels of contaminants than traditional composting. Some look to integrate vermicomposting and traditional composting in an effort to maximize efficiency while producing a high quality organic product that can be used in agriculture.

Composting food waste leads to a decrease in the quantity of greenhouse gases released into the atmosphere. In landfills, organic food waste decomposes anaerobically, producing methane gas that is emitted into the atmosphere. When this biodegradable waste is composted, it decomposes aerobically and does not produce methane, but instead produces organic compost that can then be utilized in agriculture. Recently, the city of New York has begun to require that restaurants and food-producing companies begin to compost their leftover food. Another instance of composting progress is a Wisconsin-based company called WasteCap, who is dedicated towards aiding local communities create composting plans.

Municipal Food Waste (MFW) can be composted to create this product of organic fertilizer, and many municipalities choose to do this citing environmental protection and economic efficiency as reasoning. Transporting and dumping waste in landfills requires both money and room in the landfills that have very limited available space. One municipality who chose to regulate MFW is San Francisco, who requires citizens to separate compost from trash on their own, instituting fines for non-compliance at $100 for individual homes and $500 for businesses. The city's economic reasoning for this controversial mandate is supported by their estimate that one business can save up to $30000 annually on garbage disposal costs with the implementation of the required composting. Composting is an economical and environmentally conscious step many homeowners could take to reduce their impact on landfill waste. Instead of food scraps and spoiled food taking up space in trashcans or stinking up the kitchen before the bag is full, it could be put outside and broken down by worms and added to garden beds.

Anaerobic Digestion

Anaerobic digestion produces both useful gaseous products and a solid fibrous "compostable" material. Anaerobic digestion plants can provide energy from waste by burning the methane created from food and other organic wastes to generate electricity, defraying the plants' costs and reducing greenhouse gas emissions. The United States Environmental Protection Agency states that the use of anaerobic composting allows for large amounts of food waste to avoid the landfills. Instead of producing these greenhouse gasses into the environment from being in a landfill, the gasses can alternatively be harnessed in these facilities for reuse.

Since this process of composting produces high volumes of biogas, there are potential safety issues such as explosion and poisoning. These interactions require proper maintenance and personal protective equipment is utilized. Certain U.S. states, such as Oregon, have implemented the requirement for permits on such facilities, based on the potential danger to the population and surrounding environment.

Food waste coming through the sanitary sewers from garbage disposal units is treated along with other sewage and contributes to sludge.

Commercial Liquid Food Waste

Commercially, food waste in the form of wastewater coming from commercial kitchens' sinks, dishwashers and floor drains is collected in holding tanks called grease interceptors to minimize flow to the sewer system. This often foul-smelling waste contains both organic and inorganic waste (chemical cleaners, etc.) and may also contain hazardous hydrogen sulfide gases. It is referred to as fats, oils, and grease (FOG) waste or more commonly "brown grease" (versus "yellow grease", which is fryer oil that is easily collected and processed into biodiesel) and is an overwhelming problem, especially in the US, for the aging sewer systems. Per the US EPA, sanitary sewer overflows also occur due to the improper discharge of FOGs to the collection system. Overflows discharge 3 billion US gallons (11,000,000 m3) – 10 billion US gallons (38,000,000 m3) of untreated wastewater annually into local waterways, and up to 3,700 illnesses annually are due to exposure to contamination from sanitary sewer overflows into recreational waters.

Agricultural Food Waste

Nearly all global produce, eaten or disposed of is grown using irrigated water. Irrigated water represents the largest sector of water withdraws worldwide, with as much as 90% of total water withdraws being allocated towards agricultural usage. Food which goes uneaten can account for vast quantities of water waste, with food waste being the largest area the average US citizen contributes to water waste.

SEWAGE SLUDGE

Sewage sludge is the residual, semi-solid material that is produced as a by-product during sewage treatment of industrial or municipal wastewater. The term "septage" also refers to sludge from simple wastewater treatment but is connected to simple on-site sanitation systems, such as septic tanks.

When fresh sewage or wastewater enters a primary settling tank, approximately 50% of the suspended solid matter will settle out in an hour and a half. This collection of solids is known as raw sludge or primary solids and is said to be "fresh" before anaerobic processes become active. The sludge will become putrescent in a short time once anaerobic bacteria take over, and must be removed from the sedimentation tank before this happens.

This is accomplished in one of two ways. In an Imhoff tank, fresh sludge is passed through a slot to the lower story or digestion chamber where it is decomposed by anaerobic bacteria, resulting in liquefaction and reduced volume of the sludge. After digesting for an extended period, the result is called "digested" sludge and may be disposed of by drying and then landfilling. More commonly with domestic sewage, the fresh sludge is continuously extracted from the tank mechanically and passed to separate sludge digestion tanks that operate at higher temperatures than the lower story of the Imhoff tank and, as a result, digest much more rapidly and efficiently.

"Biosolids" is a term often used in conjunction with reuse of sewage solids after sewage sludge treatment. Biosolids can be defined as organic wastewater solids that can be reused after stabilization processes such as anaerobic digestion and composting. Opponents of sewage sludge reuse reject this term as a public relations term.

Classes of Treated Sewage Sludge

- Class A sludge is typically dried and pasteurized, and is also known as "exceptional" quality.
- Class B includes all sludge not classified as Class A. Class B sludge is typically "undigested" and is volatile.

Both classes of sludge may still contain radioactive or pharmaceutical wastes.

Quantities Produced

The amount of sewage sludge produced is proportional to the amount and concentration of wastewater treated, and it also depends on the type of wastewater treatment process used. It can be expressed as kg dry solids per cubic metre of wastewater treated. The total sludge production from a wastewater treatment process is the sum of sludge from primary settling tanks (if they are part of the process configuration) plus excess sludge from the biological treatment step. For example, primary sedimentation produces about 110–170 kg/ML of so-called primary sludge, with a value of 150 kg/ML regarded as being typical for municipal wastewater in the U.S. or Europe. The sludge production is expressed as kg of dry solids produced per ML of wastewater treated; one mega litre (ML) is 103 m^3. Of the biological treatment processes, the activated sludge process produces about 70–100 kg/ML of waste activated sludge, and a trickling filter process produces slightly less sludge from the biological part of the process: 60–100 kg/ML. This means that the total sludge production of an activated sludge process that uses primary sedimentation tanks is in the range of 180–270 kg/ML, being the sum of primary sludge and waste activated sludge.

United States municipal wastewater treatment plants in 1997 produced about 7.7 million dry tons of sewage sludge, and about 6.8 million dry tons in 1998 according to EPA estimates. As of 2004, about 60% of all sewage sludge was applied to land as a soil amendment and fertilizer for growing crops.

Production of sewage sludge can be reduced by conversion from flush toilets to dry toilets such urine-diverting dry toilets and composting toilets.

Contaminants

Pathogens

Bacteria in Class A sludge products can actually regrow under certain environmental conditions. Pathogens could easily remain undetected in untreated sewage sludge. Pathogens are not a significant health issue if sewage sludge is properly treated and site-specific management practices are followed.

Micro-Pollutants

Micro-pollutants can become concentrated in sewage sludge. Each of these disposal options comes with myriad potential — and in some cases proven — human health and environment impacts.

Sterols and other hormones have been detected.

Heavy metals

One of the main concerns in the treated sludge is the concentrated metals content (arsenic, cadmium, copper, etc., some of which are also critical plant micronutrients); certain metals are regulated while others are not. Leaching methods can be used to reduce the metal content and meet the regulatory limit.

In 2009 the EPA released the Targeted National Sewage Sludge Study, which reports on the level of metals, chemicals, hormones, and other materials present in a statistical sample of sewage sludges.

Some highlights include:

- Silver is present to the degree of 20 mg/kg of sludge, on average, a near economically recoverable level, while some sludges of exceptionally high quality have up to 200 milligrams of silver per kilogram of sludge; one outlier demonstrated a silver lode of 800–900 mg per kg of sludge.
- Barium is present at the rate of 500 mg/kg, while manganese is present at the rate of 1 g/kg sludge.
- Lead, arsenic, chromium, and cadmium are estimated by the EPA to be present in detectable quantities in 100% of national sewage sludges in the US, while thallium is only estimated to be present in 94.1% of sludges.

Other Hazardous Substances

Sewage treatment plants receive various forms of hazardous waste from hospitals, nursing homes, industry and households. Low levels of constituents such as PCBs, dioxin, and brominated flame retardants, may remain in treated sludge. There are potentially thousands other components of sludge that remain untested/undetected disposed of from modern society that also end up in sludge (pharmaceuticals, nano particles, etc.) which have been proven to be hazardous to both human and ecological health.

In 2013, in South Carolina PCBs were discovered in very high levels in wastewater sludge. The problem was not discovered until thousands of acres of farm land in South Carolina were contaminated by this hazardous material. SCDHEC issued emergency regulatory order banning all PCB laden sewage sludge from being land applied on farm fields or deposited into landfills in South Carolina.

Also in 2013, after DHEC request, the city of Charlotte decided to stop land applying sewage sludge in South Carolina while authorities investigated the source of PCB contamination. In February 2014, the city of Charlotte admitted PCBs have entered their sewage treatment centers as well.

Contaminants of concern in sewage sludge are plasticizers, PDBEs, and others generated by human activities, including personal care products and medicines. Synthetic fibers from fabrics persist in treated sewage sludge as well as in biosolids-treated soils and may thus serve as an indicator of past biosolids application.

Pollutant Ceiling Concentration

The term “pollutant” is defined as part of the EPA 503 rule. The components of sludge have pollutant limits defined by the EPA. “A Pollutant is an organic substance, an inorganic substance, a combination of organic and inorganic substances, or a pathogenic organism that, after discharge and upon exposure, ingestion, inhalation, or assimilation into an organism either directly from the environment or indirectly by ingestion through the food chain, could, on the basis of information available to the Administrator of EPA, cause death, disease, behavioral abnormalities, cancer, genetic mutations, physiological malfunctions (including malfunction in reproduction), or physical deformations in either organisms or offspring of the organisms." The maximum component pollutant limits by the US EPA are:

Pollutant	Ceiling concentration (milligrams per kilogram)
Cadmium	85
Copper	4300
Lead	840
Mercury	57
Molybdenum	75
Nickel	420
Selenium	100
Zinc	7500

Treatment

Sewage sludge treatment is the process of removing contaminants from wastewater. Sewage sludge is produced from the treatment of wastewater in sewage treatment plants and consists of two basic forms: raw primary sludge and secondary sludge, also known as activated sludge in the case of the activated sludge process.

Sewage sludge is usually treated by one or several of the following treatment steps: lime stabilization, thickening, dewatering, drying, anaerobic digestion or composting. Some treatment processes, such as composting and alkaline stabilization, that involve significant amendments may affect contaminant strength and concentration: depending on the process and the contaminant in question, treatment may decrease or in some cases increase the bioavailability and solubility of contaminants.

Following treatment, sewage sludge is either landfilled, incinerated, applied on agricultural land or, in some cases, retailed or given away for free to the general public.

Health Risks

In 2011, the EPA commissioned a study at the United States National Research Council (NRC) to determine the health risks of sludge. In this document the NRC pointed out that many of the dangers of sludge are unknown and unassessed. Additionally "Regulations that limit contact with biosolids do not prevent environmental processes in the conceptual model such as aerosolization or erosion and the death or multiplication of pathogens."

The National Research Council published "Biosolids Applied to Land: Advancing Standards and Practices" in July 2002. The NRC concluded that while there is no documented scientific evidence that sewage sludge regulations have failed to protect public health, there is persistent uncertainty on possible adverse health effects. The NRC noted that further research is needed and made about 60 recommendations for addressing public health concerns, scientific uncertainties, and data gaps in the science underlying the sewage sludge standards. The EPA responded with a commitment to conduct research addressing the NRC recommendations.

Residents living near Class B sludge processing sites may experience asthma or pulmonary distress due to bioaerosols released from sludge fields.

A 2004 survey of 48 individuals near affected sites found that most reported irritation symptoms, about half reported an infection within a month of the application, and about a fourth were affected by Staphylococcus aureus, including two deaths. The number of reported S. aureus infections was 25 times as high as in hospitalized patients, a high-risk group. The authors point out that regulations call for protective gear when handling Class B biosolids and that similar protections could be considered for residents in nearby areas given the wind conditions.

Khuder, Milz, Bisesi, Vincent, McNulty, and Czajkowski conducted a health survey of persons living in close proximity to Class B sludged land. A sample of 437 people exposed to Class B sludge (living within 1-mile (1.6 km) of sludged land) - and using a control group of 176 people not exposed to sludge (not living within 1-mile (1.6 km) of sludged land) reported the following:

> Results revealed that some reported health-related symptoms were statistically significantly elevated among the exposed residents, including excessive secretion of tears, abdominal bloating, jaundice, skin ulcer, dehydration, weight loss, and general weakness. The frequency of reported occurrence of bronchitis, upper respiratory infection, and giardiasis were also statistically significantly elevated. The findings suggest an increased risk for certain respiratory, gastrointestinal, and other diseases among residents living near farm fields on which the use of biosolids was permitted.

Although correlation does not imply causation, such extensive correlations may lead reasonable people to conclude that precaution is necessary in dealing with sludge and sludged farmlands.

Harrison and Oakes suggest that, in particular, "until investigations are carried out that answer these questions (about the safety of Class B sludge), land application of Class B sludges should be viewed as a practice that subjects neighbors and workers to substantial risk of disease." They further suggest that even Class A treated sludge may have chemical contaminants (including heavy metals, such as lead) or endotoxins present, and a precautionary approach may be justified on this basis, though the vast majority of incidents reported by Lewis, et al. have been correlated with exposure to Class B untreated sludge and not Class A treated sludge.

A 2005 report by the state of North Carolina concluded that "a surveillance program of humans living near application sites should be developed to determine if there are adverse health effects in humans and animals as a result of biosolids application."

Disposal

After treatment, and dependent upon the quality of sludge produced (for example with regards to heavy metal content), sewage sludge is most commonly either disposed of in landfills, dumped in the ocean or applied to land for its fertilizing properties, as pioneered by the product Milorganite.

Ocean Dumping

It used to be common practice to dump sewage sludge into the ocean, however, this practice has stopped in many nations due to environmental concerns as well to domestic and international laws and treaties.

Land Application

Biosolids is a term widely used to denote the byproduct of domestic and commercial sewage and wastewater treatment that is to be used in agriculture. National regulations that dictate the practice of land application of treated sewage sludge differ widely and e.g. in the US there are widespread disputes about this practice.

Depending on their level of treatment and resultant pollutant content, biosolids can be used in regulated applications for non-food agriculture, food agriculture, or distribution for unlimited use. Treated biosolids can be produced in cake, granular, pellet, or liquid form and are spread over land before being incorporated into the soil or injected directly into the soil by specialist contractors. Such use was pioneered by the production of Milorganite in 1926. Use of sewage sludge has shown an increase in level of soil available phosphorus and soil salinity.

The findings of a 20-year field study of air, land, and water in Arizona, concluded that use of biosolids is sustainable and improves the soil and crops. Other studies report that plants uptake large quantities of heavy metals and toxic pollutants that are retained by produce, which is then consumed by humans.

A thesis studying the addition of sludge to neutralize soil acidity concluded that the practice was not recommended if large amounts are used because the sludge produces acids when it oxidizes.

Studies have indicated that pharmaceuticals and personal care products, which often adsorb to sludge during wastewater treatment, can persist in agricultural soils following biosolid application. Some of these chemicals, including potential endocrine disruptor Triclosan, can also travel through the soil column and leach into agricultural tile drainage at detectable levels. Other studies, however, have shown that these chemicals remain adsorbed to surface soil particles, making them more susceptible to surface erosion than infiltration. These studies are also mixed in their findings regarding the persistence of chemicals such as triclosan, triclocarban, and other pharmaceuticals. The impact of this persistence in soils is unknown, but the link to human and land animal health is likely tied to the capacity for plants to absorb and accumulate these chemicals in their consumed tissues. Studies of this kind are in early stages, but evidence of root uptake and translocation to leaves did occur for both triclosan and triclocarban in soybeans. This effect was not present in corn when tested in a different study.

A cautionary approach to land application of biosolids has been advocated by some for regions where soils have lower capacities for toxics absorption or due to the presence of unknowns in sewage biosolids. In 2007, the Northeast Regional Multi-State Research Committee (NEC 1001) issued conservative guidelines tailored to the soils and conditions typical of the northeastern US.

Use of sewage sludge is prohibited for produce to be labeled USDA-certified organic. In 2014 the United States grocery chain Whole Foods banned produce grown in sewage sludge.

Treated sewage sludge has been used in the UK, Europe and China agriculturally for more than 80 years, though there is increasing pressure in some countries to stop the practice of land application due to farm land contamination and public outrage. In the 1990s there was pressure in some European countries to ban the use of sewage sludge as a fertilizer. Switzerland, Sweden, Austria, and others introduced a ban. Since the 1960s there has been cooperative activity with industry to reduce the inputs of persistent substances from factories. This has been very successful and, for example, the content of cadmium in sewage sludge in major European cities is now only 1% of what it was in 1970.

Incineration

Sludge can also be incinerated in sludge incineration plants which comes with its own set of environmental concerns (air pollution, disposal of the ash). Pyrolysis of the sludge to create syngas and potentially biochar is possible, as is combustion of biofuel produced from drying sewage sludge or incineration in a waste-to-energy facility for direct production of electricity and steam for district heating or industrial uses.

Thermal processes can greatly reduce the volume of the sludge, as well as achieve remediation of all or some of the biological concerns. Direct waste-to-energy incineration and complete combustion systems (such as the Gate 5 Energy System) will require multi-step cleaning of the exhaust gas, to ensure no hazardous substances are released. In addition, the ash produced by incineration or incomplete combustion processes (such as fluidized-bed dryers) may be difficult to use without subsequent treatment due to high heavy metal content; solutions to this include leaching of the ashes to remove heavy metals or in the case of ash produced in a complete-combustion process, or with biochar produced from a pyrolytic process, the heavy metals may be fixed in place and the ash material readily usable as a LEEDs preferred additive to concrete or asphalt. Examples of other ways to use dried sewage sludge as an energy resource include the Gate 5 Energy System, an innovative process to power a steam turbine using heat from burning milled and dried sewage sludge, or combining dried sewage sludge with coal in coal-fired power stations. In both cases this allows for production of electricity with less carbon-dioxide emissions than conventional coal-fired power stations.

ELECTRONIC WASTE OR E-WASTE

Electronic waste or e-waste describes discarded electrical or electronic devices. Used electronics which are destined for refurbishment, reuse, resale, salvage, recycling through material recovery, or disposal are also considered e-waste. Informal processing of e-waste in developing countries can lead to adverse human health effects and environmental pollution.

Electronic scrap components, such as CPUs, contain potentially harmful materials such as lead, cadmium, beryllium, or brominated flame retardants. Recycling and disposal of e-waste may involve significant risk to health of workers and communities in developed countries and great care must be taken to avoid unsafe exposure in recycling operations and leaking of materials such as heavy metals from landfills and incinerator ashes.

Amount of Electronic Waste Worldwide

A fragment of a discarded circuit board.

Rapid changes in technology, changes in media (tapes, software, MP3), falling prices, and planned obsolescence have resulted in a fast-growing surplus of electronic waste around the globe. Technical solutions are available, but in most cases, a legal framework, a collection, logistics, and other services need to be implemented before a technical solution can be applied.

Display units (CRT, LCD, LED monitors), processors (CPU, GPU, or APU chips), memory (DRAM or SRAM), and audio components have different useful lives. Processors are most frequently out-dated (by software no longer being optimized) and are more likely to become "e-waste" while display units are most often replaced while working without repair attempts, due to changes in wealthy nation appetites for new display technology. This problem could potentially be solved with modular smartphones or Phonebloks. These types of phones are more durable and have the technology to change certain parts of the phone making them more environmentally friendly. Being able to simply replace the part of the phone that is broken will reduce e-waste. An estimated 50 million tons of e-waste are produced each year. The USA discards 30 million computers each year and 100 million phones are disposed of in Europe each year. The Environmental Protection Agency estimates that only 15–20% of e-waste is recycled, the rest of these electronics go directly into landfills and incinerators.

Electronic waste.

In 2006, the United Nations estimated the amount of worldwide electronic waste discarded each year to be 50 million metric tons. According to a report by UNEP titled, "Recycling – from E-Waste to Resources," the amount of e-waste being produced – including mobile phones and computers – could rise by as much as 500 percent over the next decade in some countries, such as India. The

United States is the world leader in producing electronic waste, tossing away about 3 million tons each year. China already produces about 2.3 million tons (2010 estimate) domestically, second only to the United States. And, despite having banned e-waste imports, China remains a major e-waste dumping ground for developed countries.

Society today revolves around technology and by the constant need for the newest and most high-tech products we are contributing to a mass amount of e-waste. Since the invention of the iPhone, cell phones have become the top source of e-waste products because they are not made to last more than two years. Electrical waste contains hazardous but also valuable and scarce materials. Up to 60 elements can be found in complex electronics. As of 2013, Apple has sold over 796 million iDevices (iPod, iPhone, iPad). Cell phone companies make cell phones that are not made to last so that the consumer will purchase new phones. Companies give these products such short lifespans because they know that the consumer will want a new product and will buy it if they make it. In the United States, an estimated 70% of heavy metals in landfills comes from discarded electronics.

While there is agreement that the number of discarded electronic devices is increasing, there is considerable disagreement about the relative risk (compared to automobile scrap, for example), and strong disagreement whether curtailing trade in used electronics will improve conditions, or make them worse. According to a study in Motherboard, attempts to restrict the trade have driven reputable companies out of the supply chain, with unintended consequences.

Global Trade Issues

Electronic waste is often exported to developing countries.

The E-waste centre of Agbogbloshie, Ghana, where electronic waste is burnt and disassembled with no safety or environmental considerations.

One theory is that increased regulation of electronic wastes and concern over the environmental harm in nature economies creates an economic disincentive to remove residues prior to export. Critics of trade in used electronics maintain that it is still too easy for brokers calling themselves recyclers to export unscreened electronic waste to developing countries, such as China, India and parts of Africa, thus avoiding the expense of removing items like bad cathode ray tubes (the processing of which is expensive and difficult). The developing countries have become toxic dump yards of e-waste. Developing countries receiving foreign e-waste often go further to repair and recycle forsaken equipment. Yet still 90% of e-waste ended up in landfills in developing countries in 2003. Proponents of international trade point to the success of fair trade programs in other industries, where cooperation has led to creation of sustainable jobs and can bring affordable technology in countries where repair and reuse rates are higher.

Rechargable batteries.

Defenders of the trade in used electronics say that extraction of metals from virgin mining has been shifted to developing countries. Recycling of copper, silver, gold, and other materials from discarded electronic devices is considered better for the environment than mining. They also state that repair and reuse of computers and televisions has become a "lost art" in wealthier nations and that refurbishing has traditionally been a path to development.

South Korea, Taiwan, and southern China all excelled in finding "retained value" in used goods, and in some cases have set up billion-dollar industries in refurbishing used ink cartridges, single-use cameras, and working CRTs. Refurbishing has traditionally been a threat to established manufacturing, and simple protectionism explains some criticism of the trade. Works like "The Waste Makers" by Vance Packard explain some of the criticism of exports of working product, for example, the ban on import of tested working Pentium 4 laptops to China, or the bans on export of used surplus working electronics by Japan.

Opponents of surplus electronics exports argue that lower environmental and labour standards, cheap labour, and the relatively high value of recovered raw materials lead to a transfer of pollution-generating activities, such as smelting of copper wire. In China, Malaysia, India, Kenya, and various African countries, electronic waste is being sent to these countries for processing, sometimes illegally. Many surplus laptops are routed to developing nations as "dumping grounds for e-waste".

Because the United States has not ratified the Basel Convention or its Ban Amendment, and has few domestic federal laws forbidding the export of toxic waste, the Basel Action Network estimates that about 80% of the electronic waste directed to recycling in the U.S. does not get recycled there at all, but is put on container ships and sent to countries such as China. This figure is disputed as an exaggeration by the EPA, the Institute of Scrap Recycling Industries, and the World Reuse, Repair and Recycling Association.

Independent research by Arizona State University showed that 87–88% of imported used computers did not have a higher value than the best value of the constituent materials they contained, and that "the official trade in end-of-life computers is thus driven by reuse as opposed to recycling".

Trade

Proponents of the trade say growth of internet access is a stronger correlation to trade than poverty. Haiti is poor and closer to the port of New York than southeast Asia, but far more electronic waste is exported from New York to Asia than to Haiti. Thousands of men, women, and children are employed in reuse, refurbishing, repair, and re-manufacturing, unsustainable industries in decline in developed countries. Denying developing nations access to used electronics may deny them sustainable employment, affordable products, and internet access, or force them to deal with even less scrupulous suppliers. In a series of seven articles for The Atlantic, Shanghai-based reporter Adam Minter describes many of these computer repair and scrap separation activities as objectively sustainable.

Sacks of waste phones in Agbogbloshie.

Opponents of the trade argue that developing countries utilize methods that are more harmful and more wasteful. An expedient and prevalent method is simply to toss equipment onto an open fire, in order to melt plastics and to burn away non-valuable metals. This releases carcinogens and neurotoxins into the air, contributing to an acrid, lingering smog. These noxious fumes include dioxins and furans. Bonfire refuse can be disposed of quickly into drainage ditches or waterways feeding the ocean or local water supplies.

In June 2008, a container of electronic waste, destined from the Port of Oakland in the U.S. to Sanshui District in mainland China, was intercepted in Hong Kong by Greenpeace. Concern over exports of electronic waste were raised in press reports in India, Ghana, Cote d'Ivoire, and Nigeria.

The research that was undertaken by the Countering WEEE Illegal Trade (CWIT) project, funded by European Commission, found that in Europe, only 35% (3.3 million tons) of all the e-waste discarded in 2012, ended up in the officially reported amounts of collection and recycling systems. The other 65% (6.15 million tons) was either:

- Exported (1.5 million tons),
- Recycled under non-compliant conditions in Europe (3.15 million tons),
- Scavenged for valuable parts (750,000 tons),
- Or simply thrown in waste bins (750,000 tons).

Guiyu

Guiyu in the Shantou region of China is a massive electronic waste processing community. It is often referred to as the "e-waste capital of the world." Traditionally, Guiyu was an agricultural community; however, in the mid-1990s it transformed into an e-waste recycling center involving over

75% of the local households and an additional 100,000 migrant workers. Thousands of individual workshops employ laborers to snip cables, pry chips from circuit boards, grind plastic computer cases into particles, and dip circuit boards in acid baths to dissolve the precious metals. Others work to strip insulation from all wiring in an attempt to salvage tiny amounts of copper wire. Uncontrolled burning, disassembly, and disposal has led to a number of environmental problems such as groundwater contamination, atmospheric pollution, and water pollution either by immediate discharge or from surface runoff (especially near coastal areas), as well as health problems including occupational safety and health effects among those directly and indirectly involved, due to the methods of processing the waste.

Six of the many villages in Guiyu specialize in circuit-board disassembly, seven in plastics and metals reprocessing, and two in wire and cable disassembly. Greenpeace, an environmental group, sampled dust, soil, river sediment, and groundwater in Guiyu. They found very high levels of toxic heavy metals and organic contaminants in both places. Lai Yun, a campaigner for the group found "over 10 poisonous metals, such as lead, mercury, and cadmium."

Guiyu is only one example of digital dumps but similar places can be found across the world in Nigeria, Ghana, and India.

Other Informal E-waste Recycling Sites

A pile of discarded TVs and computer monitors.

Guiyu is likely one of the oldest and largest informal e-waste recycling sites in the world; however, there are many sites worldwide, including India, Ghana (Agbogbloshie), Nigeria, and the Philippines. Most research involving informal e-waste recycling has been done in Guiyu, but there are a handful of studies that describe exposure levels in e-waste workers, the community, and the environment. For example, locals and migrant workers in Delhi, a northern union territory of India, scavenge discarded computer equipment and extract base metals using toxic, unsafe methods. Bangalore, located in southern India, is often referred as the "Silicon Valley of India" and has a growing informal e-waste recycling sector. A studies found that e-waste workers in the slum community had higher levels of V, Cr, Mn, Mo, Sn, Tl, and Pb than workers at an e-waste recycling facility.

Environmental Impact

Old keyboards, mouse and monitors.

The processes of dismantling and disposing of electronic waste in developing countries led to a number of environmental impacts as illustrated in the graphic. Liquid and atmospheric releases end up in bodies of water, groundwater, soil, and air and therefore in land and sea animals – both domesticated and wild, in crops eaten by both animals and human, and in drinking water.

One study of environmental effects in Guiyu, China found the following:

- Airborne dioxins – one type found at 100 times levels previously measured.
- Levels of carcinogens in duck ponds and rice paddies exceeded international standards for agricultural areas and cadmium, copper, nickel, and lead levels in rice paddies were above international standards.
- Heavy metals found in road dust – lead over 300 times that of a control village's road dust and copper over 100 times.

A separate study at the Agbogbloshie e-waste dump, Ghana found a presence of lead levels as high as 18,125 ppm in the soil. US EPA standard for lead in soil in play areas is 400 ppm and 1200 ppm for non-play areas. Scrap workers at the Agbogbloshie e-waste dump regularly burn electronic components and auto harness wires for copper recovery, releasing toxic chemicals like lead, dioxins and furans into the environment.

The environmental impact of the processing of different electronic waste components		
E-waste Component	Process Used	Potential Environmental Hazard
Cathode ray tubes (used in TVs, computer monitors, ATM, video cameras, and more)	Breaking and removal of yoke, then dumping	Lead, barium and other heavy metals leaching into the ground water and release of toxic phosphor

Printed circuit board	De-soldering and removal of computer chips; open burning and acid baths to remove metals after chips are removed.	Air emissions and discharge into rivers of glass dust, tin, lead, brominated dioxin, beryllium cadmium, and mercury
Chips and other gold plated components	Chemical stripping using nitric and hydrochloric acid and burning of chips	PAHs, heavy metals, brominated flame retardants discharged directly into rivers acidifying fish and flora. Tin and lead contamination of surface and groundwater. Air emissions of brominated dioxins, heavy metals, and PAHs
Plastics from printers, keyboards, monitors, etc.	Shredding and low temp melting to be reused	Emissions of brominated dioxins, heavy metals, and hydrocarbons
Computer wires	Open burning and stripping to remove copper	PAHs released into air, water, and soil

Information Security

E-waste presents a potential security threat to individuals and exporting countries. Hard drives that are not properly erased before the computer is disposed of can be reopened, exposing sensitive information. Credit card numbers, private financial data, account information, and records of online transactions can be accessed by most willing individuals. Organized criminals in Ghana commonly search the drives for information to use in local scams. Electronic files about government contracts have been discovered on hard drives found in Agbogbloshie. Multimillion-dollar agreements from United States security institutions such as the Defense Intelligence Agency (DIA), the Transportation Security Administration, and Homeland Security have all resurfaced in Agbogbloshie.

Recycling

Computer monitors are typically packed into low stacks on wooden pallets for recycling and then shrink-wrapped.

Recycling is an essential element of e-waste management. Properly carried out, it should greatly reduce the leakage of toxic materials into the environment and mitigate against the exhaustion of natural resources. However, it does need to be encouraged by local authorities and through community education.

One of the major challenges is recycling the printed circuit boards from the electronic wastes. The circuit boards contain such precious metals as gold, silver, platinum, etc. and such base metals as copper, iron, aluminum, etc. One way e-waste is processed is by melting circuit boards, burning cable sheathing to recover copper wire and open- pit acid leaching for separating metals of value. Conventional method employed is mechanical shredding and separation but the recycling efficiency is low. Alternative methods such as cryogenic decomposition have been studied for printed circuit board recycling, and some other methods are still under investigation. Properly disposing of or reusing electronics can help prevent health problems, reduce greenhouse-gas emissions, and create jobs. Reuse and refurbishing offer a more environmentally friendly and socially conscious alternative to downcycling processes.

Consumer Awareness Efforts

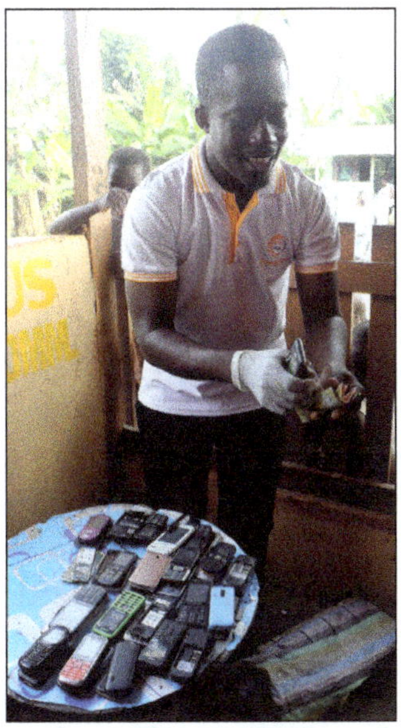

A campaign to promote e-waste recycling.

The U.S. Environmental Protection Agency encourages electronic recyclers to become certified by demonstrating to an accredited, independent third party auditor that they meet specific standards to safely recycle and manage electronics. This should work so as to ensure the highest environmental standards are being maintained. Two certifications for electronic recyclers currently exist and are endorsed by the EPA. Customers are encouraged to choose certified electronics recyclers. Responsible electronics recycling reduces environmental and human health impacts, increases the use of reusable and refurbished equipment and reduces energy use while conserving limited resources. The two EPA-endorsed certification programs are Responsible Recyclers Practices (R2) and E-Stewards. Certified companies ensure they are meeting strict environmental standards which maximize reuse and recycling, minimize exposure to human health or the environment, ensure safe management of materials and require destruction of all data used on electronics. Certified electronics recyclers have demonstrated through audits and other means that they continually meet specific high environmental standards and safely manage used electronics. Once certified, the recycler is held to the particular standard by continual oversight by the independent accredited certifying body. A certification board accredits and oversees certifying bodies to ensure that they meet specific responsibilities and are competent to audit and provide certification.

Some U.S. retailers offer opportunities for consumer recycling of discarded electronic devices. In the US, the Consumer Electronics Association (CEA) urges consumers to dispose properly of end-of-life electronics through its recycling locator. This list only includes manufacturer and retailer programs that use the strictest standards and third-party certified recycling locations, to provide consumers assurance that their products will be recycled safely and responsibly. CEA research has found that 58 percent of consumers know where to take their end-of-life electronics, and the electronics industry would very much like to see that level of awareness increase. Consumer electronics manufacturers and retailers sponsor or operate more than 5,000 recycling locations nationwide and have vowed to recycle one billion pounds annually by 2016, a sharp increase from 300 million pounds industry recycled in 2010.

The Sustainable Materials Management (SMM) Electronic Challenge was created by the United States Environmental Protection Agency (EPA) in 2012. Participants of the Challenge are manufacturers of electronics and electronic retailers. These companies collect end-of-life (EOL) electronics at various locations and send them to a certified, third-party recycler. Program participants are then able publicly promote and report 100% responsible recycling for their companies. The Electronics TakeBack Coalition (ETBC) is a campaign aimed at protecting human health and limiting environmental effects where electronics are being produced, used, and discarded. The ETBC aims to place responsibility for disposal of technology products on electronic manufacturers and brand owners, primarily through community promotions and legal enforcement initiatives. It provides recommendations for consumer recycling and a list of recyclers judged environmentally responsible. While there have been major benefits from the rise in recycling and waste collection created by producers and consumers, such as valuable materials being recovered and kept away from landfill and incineration, there are still many problems present with the EPR system including "how to ensure proper enforcement of recycling standards, what to do about waste with positive net value, and the role of competition." Many stakeholders agreed there needs to be a higher standard of accountability and efficiency to improve the systems of recycling everywhere, as well as the growing amount of waste being an opportunity more so than downfall since it gives us more chances to create an efficient system. To make recycling competition more cost-effective, the producers agreed that there needs to be a higher drive for competition because it allows them to have a wider range of producer responsibility organizations to choose from for e-waste recycling.

The Certified Electronics Recycler program for electronic recyclers is a comprehensive, integrated management system standard that incorporates key operational and continual improvement elements for quality, environmental and health and safety performance. The grassroots Silicon Valley Toxics Coalition promotes human health and addresses environmental justice problems resulting from toxins in technologies. The World Reuse, Repair, and Recycling Association is an organization dedicated to improving the quality of exported electronics, encouraging better recycling standards in importing countries, and improving practices through "Fair Trade" principles. Take Back My TV is a project of The Electronics TakeBack Coalition and grades television manufacturers to find out which are responsible, in the coalition's view, and which are not.

There have also been efforts to raise awareness of the potentially hazardous conditions of the dismantling of e-waste in American prisons. The Silicon Valley Toxics Coalition, prisoner-rights activists, and environmental groups released a Toxic Sweatshops report that details how prison labor is being used to handle e-waste, resulting in health consequences among the workers. These groups allege that, since prisons do not have adequate safety standards, inmates are dismantling the products under unhealthy and unsafe conditions.

Processing Techniques

Recycling the lead from batteries.

In many developed countries, electronic waste processing usually first involves dismantling the equipment into various parts (metal frames, power supplies, circuit boards, plastics), often by hand, but increasingly by automated shredding equipment. A typical example is the NADIN electronic waste processing plant in Novi Iskar, Bulgaria—the largest facility of its kind in Eastern Europe. The advantages of this process are the human's ability to recognize and save working and repairable parts, including chips, transistors, RAM, etc. The disadvantage is that the labor is cheapest in countries with the lowest health and safety standards.

In an alternative bulk system, a hopper conveys material for shredding into an unsophisticated mechanical separator, with screening and granulating machines to separate constituent metal and plastic fractions, which are sold to smelters or plastics recyclers. Such recycling machinery is enclosed and employs a dust collection system. Some of the emissions are caught by scrubbers and screens. Magnets, eddy currents, and Trommel screens are employed to separate glass, plastic, and ferrous and nonferrous metals, which can then be further separated at a smelter.

Leaded glass from CRTs is reused in car batteries, ammunition, and lead wheel weights, or sold to foundries as a fluxing agent in processing raw lead ore. Copper, gold, palladium, silver and tin are valuable metals sold to smelters for recycling. Hazardous smoke and gases are captured, contained and treated to mitigate environmental threat. These methods allow for safe reclamation of all valuable computer construction materials. Hewlett-Packard product recycling solutions manager Renee St. Denis describes its process as: "We move them through giant shredders about 30 feet tall and it shreds everything into pieces about the size of a quarter. Once your disk drive is shredded into pieces about this big, it's hard to get the data off". An ideal electronic waste recycling plant combines dismantling for component recovery with increased cost-effective processing of bulk electronic waste. Reuse is an alternative option to recycling because it extends the lifespan of a device. Devices still need eventual recycling, but by allowing others to purchase used electronics, recycling can be postponed and value gained from device use.

Benefits of Recycling

Recycling raw materials from end-of-life electronics is the most effective solution to the growing e-waste problem. Most electronic devices contain a variety of materials, including metals that can be recovered for future uses. By dismantling and providing reuse possibilities, intact natural resources are conserved and air and water pollution caused by hazardous disposal is avoided. Additionally, recycling reduces the amount of greenhouse gas emissions caused by the manufacturing of new products. Another benefit of recycling e-waste is that many of the materials can be recycled and re-used again. Materials that can be recycled include "ferrous (iron-based) and non-ferrous metals, glass, and various types of plastic." "Non-ferrous metals, mainly aluminum and copper can all be re-smelted and re-manufactured. Ferrous metals such as steel and iron can be also be re-used." Due to the recent surge in popularity in 3D printing, certain 3D printers have been designed (FDM variety) to produce waste that can be easily recycled which decreases the amount of harmful pollutants in the atmosphere. The excess plastic from these printers that comes out as a byproduct can also be reused to create new 3D printed creations.

Benefits of recycling are extended when responsible recycling methods are used. In the U.S., responsible recycling aims to minimize the dangers to human health and the environment that disposed and dismantled electronics can create. Responsible recycling ensures best management practices of the electronics being recycled, worker health and safety, and consideration for the environment locally and abroad. In Europe, metals that are recycled are returned to companies of origin at a reduced cost. Through a committed recycling system, manufacturers in Japan have been pushed to make their products more sustainable. Since many companies were responsible for the recycling of their own products, this imposed responsibility on manufacturers requiring many to redesign their infrastructure. As a result, manufacturers in Japan have the added option to sell the recycled metals.

Repair

One of the factors which exacerbate the e-waste problem is the diminishing lifetime of many electrical and electronic goods. There are two drivers (in particular) for this trend. On the one hand, consumer demand for low cost products mitigates against product quality and results in short product lifetimes. On the other, manufacturers in some sectors encourage a regular upgrade cycle, and may even enforce it though restricted availability of spare parts, service manuals and software updates, or through planned obsolescence.

Consumer dissatisfaction with this state of affairs has led to a growing repair movement. Often, this is at a community level such as through repair cafės or the "restart parties" promoted by the Restart Project.

The "Right to Repair" is spearheaded in the US by farmers dissatisfied with non-availability of service information, specialised tools and spare parts for their high-tech farm machinery. But the movement extends far beyond farm machinery with, for example, the restricted repair options offered by Apple coming in for criticism. Manufacturers often counter with safety concerns resulting from unauthorised repairs and modifications.

Electronic Waste Substances

Some computer components can be reused in assembling new computer products, while others are reduced to metals that can be reused in applications as varied as construction, flatware, and jewellery. Substances found in large quantities include epoxy resins, fiberglass, PCBs, PVC (polyvinyl chlorides), thermosetting plastics, lead, tin, copper, silicon, beryllium, carbon, iron, and aluminium. Elements found in small amounts include cadmium, mercury, and thallium. Elements found in trace amounts include americium, antimony, arsenic, barium, bismuth, boron, cobalt, europium, gallium, germanium, gold, indium, lithium, manganese, nickel, niobium, palladium, platinum, rhodium, ruthenium, selenium, silver, tantalum, terbium, thorium, titanium, vanadium, and yttrium. Almost all electronics contain lead and tin (as solder) and copper (as wire and printed circuit board tracks), though the use of lead-free solder is now spreading rapidly.

Several sizes of button and coin cell.

Hazardous

Recyclers in the street in São Paulo, Brazil with old computers.

- Lead

Adverse effects of lead exposure include impaired cognitive function, behavioral disturbances, attention deficits, hyperactivity, conduct problems, and lower IQ. These effects are most damaging to children whose developing nervous systems are very susceptible to damage caused by lead, cadmium, and mercury.

- Mercury

Health effects include sensory impairment, dermatitis, memory loss, and muscle weakness. Exposure in-utero causes fetal deficits in motor function, attention, and verbal domains. Environmental effects in animals include death, reduced fertility, and slower growth and development.

- Cadmium

The inhalation of cadmium can cause severe damage to the lungs and is also known to cause kidney damage. Cadmium is also associated with deficits in cognition, learning, behavior, and neuromotor skills in children.

- Hexavalent chromium

A known carcinogen after occupational inhalation exposure.

There is also evidence of cytotoxic and genotoxic effects of some chemicals, which have been shown to inhibit cell proliferation, cause cell membrane lesion, cause DNA single-strand breaks, and elevate Reactive Oxygen Species (ROS) levels.

- Sulfur

Health effects include liver damage, kidney damage, heart damage, eye and throat irritation. When released into the environment, it can create sulfuric acid through sulfur dioxide.

- Brominated Flame Retardants (BFRs)

Health effects include impaired development of the nervous system, thyroid problems, liver problems. Environmental effects: similar effects as in animals as humans. PBBs were banned from 1973 to 1977 on. PCBs were banned during the 1980s.

- Perfluorooctanoic acid (PFOA)

Studies in mice have found the following health effects: Hepatotoxicity, developmental toxicity, immunotoxicity, hormonal effects and carcinogenic effects. Studies have found increased maternal PFOA levels to be associated with an increased risk of spontaneous abortion (miscarriage) and stillbirth. Increased maternal levels of PFOA are also associated with decreases in mean gestational age (preterm birth), mean birth weight (low birth weight), mean birth length (small for gestational age), and mean APGAR score.

- Beryllium oxide

Occupational exposures associated with lung cancer, other common adverse health effects are beryllium sensitization, chronic beryllium disease, and acute beryllium disease.

- Polyvinyl chloride (PVC)

In the manufacturing phase, toxic and hazardous raw material, including dioxins are released. PVC such as chlorine tend to bioaccumulate. Over time, the compounds that contain chlorine can become pollutants in the air, water, and soil. This poses a problem as human and animals can ingest them. Additionally, exposure to toxins can result in reproductive and developmental health effects.

Generally Non-hazardous

E-waste Component	Process Used
Aluminium	Nearly all electronic goods using more than a few watts of power (heat-sinks), electrolytic capacitors.
Copper	Copper wire, printed circuit board tracks, component leads.
Germanium	1950s–1960s transistorized electronics (bipolar junction transistors).
Gold	Connector plating, primarily in computer equipment.
Lithium	Lithium-ion batteries.
Nickel	Nickel-cadmium batteries.
Silicon	Glass, transistors, ICs, printed circuit boards.
Tin	Solder, coatings on component leads.
Zinc	Plating for steel parts.

Human Health and Safety

Residents Living near the Recycling Sites

Residents living around the e-waste recycling sites, even if they do not involve in e-waste recycling activities, can also face the environmental exposure due to the food, water, and environmental contamination caused by e-waste, because they can easily contact to e-waste contaminated air, water, soil, dust, and food sources. In general, there are three main exposure pathways: inhalation, ingestion, and dermal contact.

Studies show that people living around e-waste recycling sites have a higher daily intake of heavy metals and a more serious body burden. Potential health risks include mental health, impaired cognitive function, and general physical health damage. DNA damage was also found more prevalent in all the e-waste exposed populations (i.e. adults, children, and neonates) than the populations in the control area. DNA breaks can increase the likelihood of wrong replication and thus mutation, as well as lead to cancer if the damage is to a tumor suppressor gene.

Prenatal Exposure and Neonates' Health

Prenatal exposure to e-waste has found to have adverse effects on human body burden of pollutants of the neonates. In Guiyu, one of the most famous e-waste recycling sites in China, it was found that increased cord blood lead concentration of neonates was associated with parents' participation in e-waste recycling processes, as well as how long the mothers spent living in Guiyu and in e-waste recycling factories or workshops during pregnancy. Besides, a higher placental metallothionein (a small protein marking the exposure of toxic metals) was found among neonates from Guiyu as a result of Cd exposure, while the higher Cd level in Guiyu's neonates was related to the involvement in e-waste recycling of their parents. High PFOA exposure of mothers in Guiyu is related to adverse effect on growth of their new-born and the prepotency in this area.

Prenatal exposure to informal e-waste recycling can also lead to several adverse birth outcomes (still birth, low birth weight, low Apgar scores, etc.) and longterm effects such as behavioral and learning problems of the neonates in their future life.

Children

Children are especially sensitive to e-waste exposure because of several reasons, such as their smaller size, higher metabolism rate, larger surface area in relation to their weight, and multiple exposure pathways (for example, dermal, hand-to-mouth, and take-home exposure). They were measured to have an 8-time potential health risk compared to the adult e-waste recycling workers. Studies have found significant higher blood lead levels (BLL) and blood cadmium levels (BCL) of children living in e-waste recycling area compared to those living in control area. For example, one study found that the average BLL in Guiyu was nearly 1.5 times compared to that in the control site (15.3 ug/dL compared to 9.9 ug/dL), while the CDC of the United States has set a reference level for blood lead at 5 ug/dL. The highest concentrations of lead were found in the children of parents whose workshop dealt with circuit boards and the lowest was among those who recycled plastic.

Exposure to e-waste can cause serious health problems to children. Children's exposure to developmental neurotoxins containing in e-waste such as lead, mercury, cadmium, chromium and PBDEs can lead to a higher risk of lower IQ, impaired cognitive function, and other adverse effects. In certain age groups, a decreased lung function of children in e-waste recycling sites has been found. Some studies also found associations between children's e-waste exposure and impaired coagulation, hearing loss, and decreased vaccine antibody tilters in e-waste recycling area.

E-waste Recycling Workers

The Occupational Safety & Health Administration (OSHA) has summarized several potential safety hazards of recycling workers in general, such as crushing hazards, hazardous energy released, and toxic metals.

Hazards applicable to recycling in general	
Hazards	Details
Slips, trips, and falls	They can happen during collecting and transporting e-wastes.
Crushing hazards	Workers can be stuck or crushed by the machine or the e-waste. There can be traffic accidents when transporting e-waste. Using machines that have moving parts, such as conveyors and rolling machines can also cause crush accidents, leading to amputations, crushed fingers or hands.
Hazardous energy released	Unexpected machine startup can cause death or injury to workers. This can happen during the installation, maintenance, or repair of machines, equipment, processes, or systems.
Cuts and lacerations	Hands or body injuries and eye injuries can occur when dismantling e-wastes that have sharp edges.
Noise	Working overtime near loud noises from drilling, hammering, and other tools that can make a great noise lead to hearing loss.
Toxic chemicals (dusts)	Burning e-waste to extract metals emits toxic chemicals (e.g. PAHs, lead) from e-waste to the air, which can be inhaled or ingested by workers at recycling sites. This can lead to illness from toxic chemicals.

OSHA has also specified some chemical components of electronics that can potentially do harm to e-recycling workers' health, such as lead, mercury, PCBs, asbestos, refractory ceramic fibers (RCFs), and radioactive substances.Besides, in the United States, most of these chemical hazards have specific Occupational exposure limits (OELs) set by OSHA, National Institute for Occupational Safety and Health (NIOSH), and American Conference of Governmental Industrial Hygienists (ACGIH).

Occupational exposure limits (OELs) of some hazardous chemicals		
Hazardous chemicals	OELs (mg/m^3)	Type of OELs
lead (Pb)	0.05	NIOSH recommended exposure limits (REL), time weighted average (TWA)
mercury (Hg)	0.05	NIOSH REL, TWA
cadmium (Cd)	0.005	OSHA permissible exposure limit (PEL), TWA
hexavalent chromium	0.005	OSHA PEL, TWA
sulfer dioxide	5	NIOSH REL, TWA

Informal and Formal Industries

Informal e-recycling industry refers to small e-waste recycling workshops with few (if any) automatic procedures and personal protective equipment (PPE). On the other hand, formal e-recycling industry refers to regular e-recycling facilities sorting materials from e-waste with automatic machinery and manual labor, where pollution control and PPE are common. Sometimes formal e-recycling facilities dismantle the e-waste to sort materials, then distribute it to other downstream recycling department to further recover materials such as plastic and metals.

The health impact of e-waste recycling workers working in informal industry and formal industry are expect to be different in the extent. Studies in three recycling sites in China suggest that the health risks of workers from formal e-recycling facilities in Jiangsu and Shanghai were lower compared to those worked in informal e-recycling sites in Guiyu. In another study of e-waste recycling in India, hair samples were collected from workers at an e-waste recycling facility and an e-waste recycling slum community (informal industry) in Bangalore. Levels of V, Cr, Mn, Mo, Sn, Tl, and Pb were significantly higher in the workers at the e-waste recycling facility compared to the e-waste workers in the slum community. However, Co, Ag, Cd, and Hg levels were significantly higher in the slum community workers compared to the facility workers.

Even in formal e-recycling industry, workers can be exposed to excessive pollutants. Studies in the formal e-recycling facilities in France and Sweden found workers' overexposure (compared to recommended occupational guidelines) to lead, cadmium, mercury and some other metals, as well as BFRs, PCBs, dioxin and furans. Workers in formal industry are also exposed to more brominated flame-retardants than reference groups.

Hazard Controls

For occupational health and safety of e-waste recycling workers, both employers and workers should take actions. Suggestions for the e-waste facility employers and workers given by California Department of Public Health are illustrated in the graphic.

Safety suggestion for e-waste recycling facilities employers and workers		
Hazards	What must employers do	What should workers do
General	Actions include: 1. determine the hazards in the workplace and take actions to control them; 2. check and make correction to the workplace condition regularly; 3. supply safe tools and PPE to workers; 4. provide workers with training about hazards and safe work practice; 5. a written document about injury and illness prevention.	Suggestions include: 1. wear PPE when working; 2. talk with employers about ways to improve working conditions; 3. report anything unsafe in the workplace to employers; 4. share experience of how to work safely with new workers.
Dust	Actions include: 1. offer a clean eating area, cleaning area and supplies, uniforms and shoes, and lockers for clean clothes to the workers; 2. provide tools to dismantle the e-waste. If the dust contains lead or cadmium: 1. measure the dust, lead and cadmium level in the air; 2. provide cleaning facilities such as wet mops and vacuums; 3. provide exhaust ventilation. If it is still not sufficient to reduce the dust, provide workers with respirators; 4. provide workers with blood lead testing when lead level is not less than 30 mg/m^3.	Protective measures include: 1. clean the workplace regularly, and do not eat or smoke when dealing with e-waste; 2. don't use brooms to clean the workplace since brooms can raise dust; 3. before going home, shower, change into clean clothes, and separate the dirty work clothes and clean clothes; 4. test the blood lead, even if the employers don't provide it; 5. use respirator, check for leaks every time before use, always keep it on your face in the respirator use area, and clean it properly after use.

Cuts and lacerations	Protective equipment such as gloves, masks and eye protection equipments should be provided to workers.	When dealing with glass or shredding materials, protect the hands and arms using special gloves and oversleeves.
Noise	Actions include: 1. measure the noise in the workplace, and use engineering controls when levels exceed the exposure limit; 2. reduce the vibration of the working desk by rubber matting; 3. provide workers with earmuffs when necessary.	Wear the hearing protection all the time when working. Ask for the employer about the noise monitoring results. Test the hearing ability.
Lifting injuries	Provide facilities to lift or move the e-waste and adjustable work tables.	When handling e-waste, try to decrease the load per time. Try to get help from other workers when lifting heavy or big things.

INDUSTRIAL WASTE

We have millions of factories, mills, industries, mining plants, etc. around the world. These industries use raw materials to produce finished goods for consumers. But in the manufacturing process, there are materials which are rendered useless. They constitute the industrial waste. Some examples of industrial wastes are metals, paints, sandpaper, slag, ash, radioactive wastes, etc. In this topic, we will discuss different types of industrial wastes and their disposal.

Types of Industrial Waste

Industrial waste can be categorized into biodegradable and non-biodegradable.

Biodegradable: Those industrial wastes which can be decomposed into the non-poisonous matter by the action of certain microorganisms are the biodegradable wastes. They are even comparable to house wastes. These kinds of waste are generated from food processing industries, dairy, textile mills, slaughterhouses, etc. Some examples are paper, leather, wool, animal bones, wheat, etc. They are not toxic in nature and they do not require special treatment either. Their treatment processes include combustion, composting, gasification, bio-methanation, etc.

Non-biodegradable: Those industrial wastes which cannot be decomposed into non-poisonous substances are the non-biodegradable wastes. Examples are plastics, fly ash, synthetic fibers, gypsum, silver foil, glass objects, radioactive wastes, etc. They are generated by iron and steel plants, fertilizer industries, chemical, drugs, and dyes industries. It is estimated that about 10 to 15 percentage of the total industrial wastes are non-biodegradable and hazardous, and the rate of increase in this category of waste is only increasing every year. These wastes cannot be broken down easily and made less harmful. Hence, they pollute the environment and cause threat to living organisms. They accumulate in the environment and enter the bodies of animals and plants causing diseases. However, with the advancement in technology, several disposals, and reuse methods have been developed. Wastes from one industry are being treated and utilized in another industry. For example, the cement industry uses the slag and fly ash generated as waste by steel industries. Landfill and incineration are other methods which are being resorted to, for the treated of hazardous wastes.

RADIOACTIVE WASTE

Radioactive waste is waste that contains radioactive material. Radioactive waste is usually a by-product of nuclear power generation and other applications of nuclear fission or nuclear technology, such as research and medicine. Radioactive waste is hazardous to most forms of life and the environment, and is regulated by government agencies in order to protect human health and the environment.

Radioactivity naturally decays over time, so radioactive waste has to be isolated and confined in appropriate disposal facilities for a sufficient period until it no longer poses a threat. The time radioactive waste must be stored for depends on the type of waste and radioactive isotopes. Current approaches to managing radioactive waste have been segregation and storage for short-lived waste, near-surface disposal for low and some intermediate level waste, and deep burial or partitioning/transmutation for the high-level waste.

TINT low-level radioactive waste barrels.

A summary of the amounts of radioactive waste and management approaches for most developed countries are presented and reviewed periodically as part of the International Atomic Energy Agency (IAEA) Joint Convention on the Safety of Spent Fuel Management and on the Safety of Radioactive Waste Management.

Sources

Radioactive waste comes from a number of sources. In countries with nuclear power plants, nuclear armament, or nuclear fuel treatment plants, the majority of waste originates from the nuclear fuel cycle and nuclear weapons reprocessing. Other sources include medical and industrial wastes, as well as naturally occurring radioactive materials (NORM) that can be concentrated as a result of the processing or consumption of coal, oil and gas, and some minerals, as discussed below.

Nuclear Fuel Cycle

Front End

Waste from the front end of the nuclear fuel cycle is usually alpha-emitting waste from the extraction of uranium. It often contains radium and its decay products.

Uranium dioxide (UO_2) concentrate from mining is a thousand or so times as radioactive as the granite used in buildings. It is refined from yellowcake (U_3O_8), then converted to uranium hexafluoride gas (UF_6). As a gas, it undergoes enrichment to increase the U-235 content from 0.7% to about 4.4% (LEU). It is then turned into a hard ceramic oxide (UO_2) for assembly as reactor fuel elements.

The main by-product of enrichment is depleted uranium (DU), principally the U-238 isotope, with a U-235 content of ~0.3%. It is stored, either as UF_6 or as U_3O_8. Some is used in applications where its extremely high density makes it valuable such as anti-tank shells, and on at least one occasion even a sailboat keel. It is also used with plutonium for making mixed oxide fuel (MOX) and to dilute, or downblend, highly enriched uranium from weapons stockpiles which is now being redirected to become reactor fuel.

Back End

The back-end of the nuclear fuel cycle, mostly spent fuel rods, contains fission products that emit beta and gamma radiation, and actinides that emit alpha particles, such as uranium-234 (half-life 245 thousand years), neptunium-237 (2.144 million years), plutonium-238 (87.7 years) and americium-241 (432 years), and even sometimes some neutron emitters such as californium (half-life of 898 years for Cf-251). These isotopes are formed in nuclear reactors.

It is important to distinguish the processing of uranium to make fuel from the reprocessing of used fuel. Used fuel contains the highly radioactive products of fission. Many of these are neutron absorbers, called neutron poisons in this context. These eventually build up to a level where they absorb so many neutrons that the chain reaction stops, even with the control rods completely removed. At that point the fuel has to be replaced in the reactor with fresh fuel, even though there is still a substantial quantity of uranium-235 and plutonium present. In the United States, this used fuel is usually "stored", while in other countries such as Russia, the United Kingdom, France, Japan and India, the fuel is reprocessed to remove the fission products, and the fuel can then be re-used. The fission products removed from the fuel are a concentrated form of high-level waste as are the chemicals used in the process. While these countries reprocess the fuel carrying out single plutonium cycles, India is the only country known to be planning multiple plutonium recycling schemes.

Fuel composition and long term radioactivity

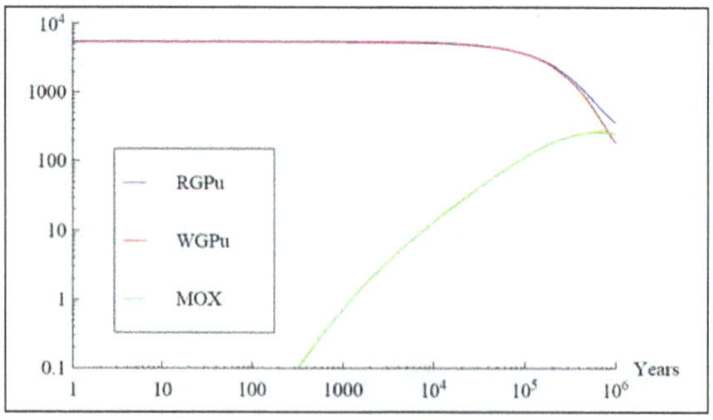

Activity of U-233 for three fuel types. In the case of MOX, the U-233 increases for the first 650 thousand yearsas it is produced by decay of Np-237 which was created in the reactor by absorption of neutrons by U-235.

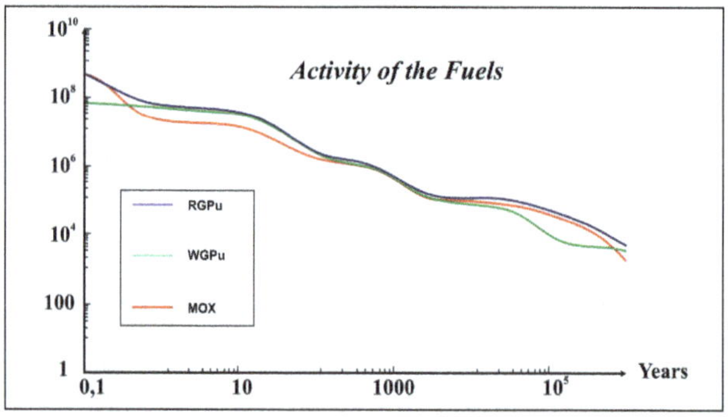

Total activity for three fuel types. In region 1 we have radiation from short-lived nuclides, and in region 2 from Sr-90 and Cs-137. On the far right we see the decay of Np-237 and U-233.

The use of different fuels in nuclear reactors results in different spent nuclear fuel (SNF) composition, with varying activity curves.

Long-lived radioactive waste from the back end of the fuel cycle is especially relevant when designing a complete waste management plan for SNF. When looking at long-term radioactive decay, the actinides in the SNF have a significant influence due to their characteristically long half-lives. Depending on what a nuclear reactor is fueled with, the actinide composition in the SNF will be different.

An example of this effect is the use of nuclear fuels with thorium. Th-232 is a fertile material that can undergo a neutron capture reaction and two beta minus decays, resulting in the production of fissile U-233. The SNF of a cycle with thorium will contain U-233. Its radioactive decay will strongly influence the long-term activity curve of the SNF around a million years. A comparison of the activity associated to U-233 for three different SNF types can be seen in the figure above. The burnt fuels are thorium with reactor-grade plutonium (RGPu), thorium with weapons-grade plutonium (WGPu) and Mixed Oxide fuel (MOX, no thorium). For RGPu and WGPu, the initial amount of U-233 and its decay around a million years can be seen. This has an effect in the total activity curve of the three fuel types. The initial absence of U-233 and its daughter products in the MOX fuel results in a lower activity in region 3 of the figure on the bottom right, whereas for RGPu and WGPu the curve is maintained higher due to the presence of U-233 that has not fully decayed. Nuclear reprocessing can remove the actinides from the spent fuel so they can be used or destroyed.

Proliferation Concerns

Since uranium and plutonium are nuclear weapons materials, there have been proliferation concerns. Ordinarily (in spent nuclear fuel), plutonium is reactor-grade plutonium. In addition to plutonium-239, which is highly suitable for building nuclear weapons, it contains large amounts of undesirable contaminants: plutonium-240, plutonium-241, and plutonium-238. These isotopes are extremely difficult to separate, and more cost-effective ways of obtaining fissile material exist (e.g., uranium enrichment or dedicated plutonium production reactors).

High-level waste is full of highly radioactive fission products, most of which are relatively short-lived. This is a concern since if the waste is stored, perhaps in deep geological storage, over many years the fission products decay, decreasing the radioactivity of the waste and making the plutonium easier to access. The undesirable contaminant Pu-240 decays faster than the Pu-239, and thus the quality of the bomb material increases with time (although its quantity decreases during that time as well). Thus, some have argued, as time passes, these deep storage areas have the potential to become "plutonium mines", from which material for nuclear weapons can be acquired with relatively little difficulty. Critics of the latter idea have pointed out the difficulty of recovering useful material from sealed deep storage areas makes other methods preferable. Specifically, the high radioactivity and heat (80 °C in surrounding rock) greatly increases the difficulty of mining a storage area, and the enrichment methods required have high capital costs.

Pu-239 decays to U-235 which is suitable for weapons and which has a very long half-life (roughly 109 years). Thus plutonium may decay and leave uranium-235. However, modern reactors are only moderately enriched with U-235 relative to U-238, so the U-238 continues to serve as a denaturation agent for any U-235 produced by plutonium decay.

One solution to this problem is to recycle the plutonium and use it as a fuel e.g. in fast reactors. In pyrometallurgical fast reactors, the separated plutonium and uranium are contaminated by actinides and cannot be used for nuclear weapons.

Nuclear Weapons Decommissioning

Waste from nuclear weapons decommissioning is unlikely to contain much beta or gamma activity other than tritium and americium. It is more likely to contain alpha-emitting actinides such as Pu-239 which is a fissile material used in bombs, plus some material with much higher specific activities, such as Pu-238 or Po.

In the past the neutron trigger for an atomic bomb tended to be beryllium and a high activity alpha emitter such as polonium; an alternative to polonium is Pu-238. For reasons of national security, details of the design of modern bombs are normally not released to the open literature.

Some designs might contain a radioisotope thermoelectric generator using Pu-238 to provide a long lasting source of electrical power for the electronics in the device.

It is likely that the fissile material of an old bomb which is due for refitting will contain decay products of the plutonium isotopes used in it, these are likely to include U-236 from Pu-240 impurities, plus some U-235 from decay of the Pu-239; due to the relatively long half-life of these Pu isotopes, these wastes from radioactive decay of bomb core material would be very small, and in any case, far less dangerous (even in terms of simple radioactivity) than the Pu-239 itself.

The beta decay of Pu-241 forms Am-241; the in-growth of americium is likely to be a greater problem than the decay of Pu-239 and Pu-240 as the americium is a gamma emitter (increasing external-exposure to workers) and is an alpha emitter which can cause the generation of heat. The plutonium could be separated from the americium by several different processes; these would include pyrochemical processes and aqueous/organic solvent extraction. A truncated PUREX type extraction process would be one possible method of making the separation. Naturally occurring uranium is not fissile because it contains 99.3% of U-238 and only 0.7% of U-235.

Legacy Waste

Due to historic activities typically related to radium industry, uranium mining, and military programs, numerous sites contain or are contaminated with radioactivity. In the United States alone, the Department of Energy states there are "millions of gallons of radioactive waste" as well as "thousands of tons of spent nuclear fuel and material" and also "huge quantities of contaminated soil and water." Despite copious quantities of waste, the DOE has stated a goal of cleaning all presently contaminated sites successfully by 2025. The Fernald, Ohio site for example had "31 million pounds of uranium product", "2.5 billion pounds of waste", "2.75 million cubic yards of contaminated soil and debris", and a "223 acre portion of the underlying Great Miami Aquifer had uranium levels above drinking standards." The United States has at least 108 sites designated as areas that are contaminated and unusable, sometimes many thousands of acres. DOE wishes to clean or mitigate many or all by 2025, using the recently developed method of geomelting, however the task can be difficult and it acknowledges that some may never be completely remediated. In just one of these 108 larger designations, Oak Ridge National Laboratory, there were for example at least "167 known contaminant release sites" in one of the three subdivisions of the 37,000-acre (150 km^2) site. Some of the U.S. sites were smaller in nature, however, cleanup issues were simpler to address, and DOE has successfully completed cleanup, or at least closure, of several sites.

Medicine

Radioactive medical waste tends to contain beta particle and gamma ray emitters. It can be divided into two main classes. In diagnostic nuclear medicine a number of short-lived gamma emitters such as technetium-99m are used. Many of these can be disposed of by leaving it to decay for a short time before disposal as normal waste. Other isotopes used in medicine, with half-lives in parentheses, include:

- Y-90, used for treating lymphoma (2.7 days).
- I-131, used for thyroid function tests and for treating thyroid cancer (8.0 days).
- Sr-89, used for treating bone cancer, intravenous injection (52 days).
- Ir-192, used for brachytherapy (74 days).
- Co-60, used for brachytherapy and external radiotherapy (5.3 years).
- Cs-137, used for brachytherapy, external radiotherapy (30 years).

Industry

Industrial source waste can contain alpha, beta, neutron or gamma emitters. Gamma emitters are used in radiography while neutron emitting sources are used in a range of applications, such as oil well logging.

Naturally Occurring Radioactive Material

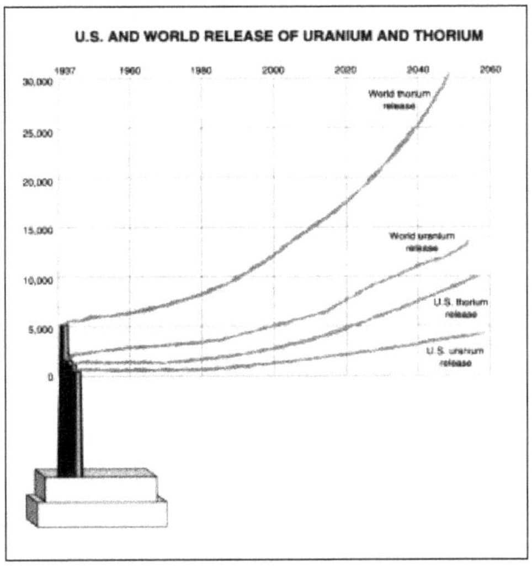

Annual release of uranium and thorium radioisotopes from coal combustion, predicted by ORNL to cumulatively amount to 2.9 million tons over the 1937–2040 period, from the combustion of an estimated 637 billion tons of coal worldwide.

Substances containing natural radioactivity are known as NORM (Naturally occurring radioactive material). After human processing that exposes or concentrates this natural radioactivity (such as mining bringing coal to the surface or burning it to produce concentrated ash), it becomes technologically enhanced naturally occurring radioactive material (TENORM). A lot of this waste is alpha particle-emitting matter from the decay chains of uranium and thorium. The main source of radiation in the human body is potassium-40 (40K), typically 17 milligrams in the body at a time and 0.4 milligrams/day intake. Most rocks, due to their components, have a low level of radioactivity. Usually ranging from 1 millisievert (mSv) to 13 mSv annually depending on location, average radiation exposure from natural radioisotopes is 2.0 mSv per person a year worldwide. This makes up the majority of typical total dosage (with mean annual exposure from other sources amounting to 0.6 mSv from medical tests averaged over the whole populace, 0.4 mSv from cosmic rays, 0.005 mSv from the legacy of past atmospheric nuclear testing, 0.005 mSv occupational exposure, 0.002 mSv from the Chernobyl disaster, and 0.0002 mSv from the nuclear fuel cycle).

TENORM is not regulated as restrictively as nuclear reactor waste, though there are no significant differences in the radiological risks of these materials.

Coal

Coal contains a small amount of radioactive uranium, barium, thorium and potassium, but, in the case of pure coal, this is significantly less than the average concentration of those elements in the Earth's crust. The surrounding strata, if shale or mudstone, often contain slightly more than average and this may also be reflected in the ash content of 'dirty' coals. The more active ash minerals become concentrated in the fly ash precisely because they do not burn well. The radioactivity of fly ash is about the same as black shale and is less than phosphate rocks, but is more of a concern because a small amount of the fly ash ends up in the atmosphere where it can be inhaled. According to U.S. NCRP reports, population exposure from 1000-MWe power plants amounts to 490 person-rem/year for coal power plants, 100 times as great as nuclear power plants (4.8 person-rem/year). (The exposure from the complete nuclear fuel cycle from mining to waste disposal is 136 person-rem/year; the corresponding value for coal use from mining to waste disposal is "probably unknown").

Oil and Gas

Residues from the oil and gas industry often contain radium and its decay products. The sulfate scale from an oil well can be very radium rich, while the water, oil and gas from a well often contain radon. The radon decays to form solid radioisotopes which form coatings on the inside of pipework. In an oil processing plant the area of the plant where propane is processed is often one of the more contaminated areas of the plant as radon has a similar boiling point to propane.

Classification

Classifications of radioactive waste varies by country. The IAEA, which publishes the Radioactive Waste Safety Standards (RADWASS), also plays a significant role.

Mill Tailings

Removal of very low-level waste.

Uranium tailings are waste by-product materials left over from the rough processing of uranium-bearing ore. They are not significantly radioactive. Mill tailings are sometimes referred to as 11(e)2 wastes, from the section of the Atomic Energy Act of 1946 that defines them. Uranium mill tailings typically also contain chemically hazardous heavy metal such as lead and arsenic. Vast mounds of uranium mill tailings are left at many old mining sites, especially in Colorado, New Mexico, and Utah.

Although mill tailings are not very radioactive, they have long half-lives. Mill tailings often contain radium, thorium and trace amounts of uranium.

Low-level Waste

Low level waste (LLW) is generated from hospitals and industry, as well as the nuclear fuel cycle. Low-level wastes include paper, rags, tools, clothing, filters, and other materials which contain small amounts of mostly short-lived radioactivity. Materials that originate from any region of an active area are commonly designated as LLW as a precautionary measure even if there is only a remote possibility of being contaminated with radioactive materials. Such LLW typically exhibits no higher radioactivity than one would expect from the same material disposed of in a non-active area, such as a normal office block. Example LLW includes wiping rags, mops, medical tubes, laboratory animal carcasses, and more.

Some high-activity LLW requires shielding during handling and transport but most LLW is suitable for shallow land burial. To reduce its volume, it is often compacted or incinerated before disposal. Low-level waste is divided into four classes: class A, class B, class C, and Greater Than Class C (GTCC).

Intermediate-level Waste

Spent fuel flasks are transported by railway. Each flask is constructed of 14 in (360 mm) thick solid steel and weighs in excess of 50 tons.

Intermediate-level waste (ILW) contains higher amounts of radioactivity and in general require shielding, but not cooling. Intermediate-level wastes includes resins, chemical sludge and metal nuclear fuel cladding, as well as contaminated materials from reactor decommissioning. It may be solidified in concrete or bitumen for disposal. As a general rule, short-lived waste (mainly non-fuel materials from reactors) is buried in shallow repositories, while long-lived waste (from fuel and fuel reprocessing) is deposited in geological repository. U.S. regulations do not define this category of waste; the term is used in Europe and elsewhere.

High-level Waste

High-level waste (HLW) is produced by nuclear reactors. The exact definition of HLW differs internationally. After a nuclear fuel rod serves one fuel cycle and is removed from the core, it is considered HLW. Fuel rods contain fission products and transuranic elements generated in the reactor core. Spent fuel is highly radioactive and often hot. HLW accounts for over 95 percent of the total radioactivity produced in the process of nuclear electricity generation. The amount of HLW worldwide is currently increasing by about 12,000 metric tons every year, which is the equivalent to about 100 double-decker buses (~200 single-decker buses) or a two-story structure with a footprint the size of a basketball court. A 1000-MW nuclear power plant produces about 27 tonnes of spent nuclear fuel (unreprocessed) every year. In 2010, there was very roughly estimated to be stored some 250,000 tons of nuclear HLW, that does not include amounts that have escaped into the environment from accidents or tests. Japan estimated to hold 17,000 tons of HLW in storage in 2015. HLW have been shipped to other countries to be stored or reprocessed, and in some cases, shipped back as active fuel.

The radioactive waste from spent fuel rods consist primarily of cesium-137 and strontium-90, but it may also include plutonium, which can be considered a transuranic waste. The half-lives of these radioactive elements can differ quite extremely. Some elements, such as cesium-137 and strontium-90 have half-lives of approximately 30 years. Meanwhile, plutonium has a half-life of that can stretch to as long as 24,000 years.

The ongoing controversy over high-level radioactive waste disposal is a major constraint on the nuclear power's global expansion. Most scientists agree that the main proposed long-term solution is deep geological burial, either in a mine or a deep borehole. However, almost six decades after commercial nuclear energy began, no government has succeeded in opening such a repository for civilian high-level nuclear waste, although Finland is in the advanced stage of the construction of such facility, the Onkalo spent nuclear fuel repository. Reprocessing or recycling spent nuclear fuel options already available or under active development still generate waste and so are not a total solution, but can reduce the sheer quantity of waste, and there are many such active programs worldwide. Deep geological burial remains the only responsible way to deal with high-level nuclear waste. The Morris Operation is currently the only de facto high-level radioactive waste storage site in the United States.

Transuranic Waste

Transuranic waste (TRUW) as defined by U.S. regulations is, without regard to form or origin, waste that is contaminated with alpha-emitting transuranic radionuclides with half-lives greater than 20 years and concentrations greater than 100 nCi/g (3.7 MBq/kg), excluding high-level waste. Elements that have an atomic number greater than uranium are called transuranic ("beyond uranium"). Because of their long half-lives, TRUW is disposed more cautiously than either low- or intermediate-level waste. In the U.S., it arises mainly from weapons production, and consists of clothing, tools, rags, residues, debris and other items contaminated with small amounts of radioactive elements (mainly plutonium).

Under U.S. law, transuranic waste is further categorized into "contact-handled" (CH) and "remote-handled" (RH) on the basis of the radiation dose rate measured at the surface of the waste container. CH TRUW has a surface dose rate not greater than 200 mrem per hour (2 mSv/h), whereas RH TRUW has a surface dose rate of 200 mrem/h (2 mSv/h) or greater. CH TRUW does not have the very high radioactivity of high-level waste, nor its high heat generation, but RH TRUW can be highly radioactive, with surface dose rates up to 1,000,000 mrem/h (10,000 mSv/h). The U.S. currently disposes of TRUW generated from military facilities at the Waste Isolation Pilot Plant (WIPP) in a deep salt formation in New Mexico.

Prevention

A theoretical way to reduce waste accumulation is to phase out current reactors in favour of Generation IV Reactors, which output less waste per power generated. Fast reactors can theoretically consume some existing waste. The UK's Nuclear Decommissioning Authority published a position paper in 2014 on the progress on approaches to the management of separated plutonium, which summarises the conclusions of the work that NDA shared with UK government.

Management

Modern medium to high level transport container for nuclear waste.

Of particular concern in nuclear waste management are two long-lived fission products, Tc-99 (half-life 220,000 years) and I-129 (half-life 15.7 million years), which dominate spent fuel radioactivity after a few thousand years. The most troublesome transuranic elements in spent fuel are Np-237 (half-life two million years) and Pu-239 (half-life 24,000 years). Nuclear waste requires sophisticated treatment and management to successfully isolate it from interacting with the biosphere. This usually necessitates treatment, followed by a long-term management strategy involving storage, disposal or transformation of the waste into a non-toxic form. Governments around the world are considering a range of waste management and disposal options, though there has been limited progress toward long-term waste management solutions.

In the second half of the 20th century, several methods of disposal of radioactive waste were investigated by nuclear nations, which are :

- "Long term above ground storage", not implemented.
- "Disposal in outer space" (for instance, inside the Sun), not implemented - as it would be currently too expensive.
- "Deep borehole disposal", not implemented.
- "Rock-melting", not implemented.
- "Disposal at subduction zones", not implemented.

- "Ocean disposal", used to be done by the USSR, the United Kingdom, Switzerland, the United States, Belgium, France, The Netherlands, Japan, Sweden, Russia, Germany, Italy and South Korea. This is no longer permitted by international agreements.
- "Sub seabed disposal", not implemented, not permitted by international agreements.
- "Disposal in ice sheets", rejected in Antarctic Treaty.
- "Direct injection", done by USSR and USA.
- Nuclear transmutation, using lasers to cause beta decay to convert the unstable atoms to those with shorter half-lives.

In the US, waste management policy completely broke down with the ending of work on the incomplete Yucca Mountain Repository. At present there are 70 nuclear power plant sites where spent fuel is stored. A Blue Ribbon Commission was appointed by President Obama to look into future options for this and future waste. A deep geological repository seems to be favored.

Initial Treatment

Vitrification

Long-term storage of radioactive waste requires the stabilization of the waste into a form which will neither react nor degrade for extended periods. It is theorized that one way to do this might be through vitrification. Currently at Sellafield the high-level waste (PUREX first cycle raffinate) is mixed with sugar and then calcined. Calcination involves passing the waste through a heated, rotating tube. The purposes of calcination are to evaporate the water from the waste, and de-nitrate the fission products to assist the stability of the glass produced.

The 'calcine' generated is fed continuously into an induction heated furnace with fragmented glass. The resulting glass is a new substance in which the waste products are bonded into the glass matrix when it solidifies. As a melt, this product is poured into stainless steel cylindrical containers ("cylinders") in a batch process. When cooled, the fluid solidifies ("vitrifies") into the glass. After being formed, the glass is highly resistant to water.

After filling a cylinder, a seal is welded onto the cylinder head. The cylinder is then washed. After being inspected for external contamination, the steel cylinder is stored, usually in an underground repository. In this form, the waste products are expected to be immobilized for thousands of years.

The glass inside a cylinder is usually a black glossy substance. All this work (in the United Kingdom) is done using hot cell systems. Sugar is added to control the ruthenium chemistry and to stop the formation of the volatile RuO_4 containing radioactive ruthenium isotopes. In the West, the glass is normally a borosilicate glass (similar to Pyrex), while in the former Soviet bloc it is normal to use a phosphate glass. The amount of fission products in the glass must be limited because some (palladium, the other Pt group metals, and tellurium) tend to form metallic phases which separate from the glass. Bulk vitrification uses electrodes to melt soil and wastes, which are then buried underground. In Germany a vitrification plant is in use; this is treating the waste from a small demonstration reprocessing plant which has since been closed down.

Ion Exchange

It is common for medium active wastes in the nuclear industry to be treated with ion exchange or other means to concentrate the radioactivity into a small volume. The much less radioactive bulk (after treatment) is often then discharged. For instance, it is possible to use a ferric hydroxide floc to remove radioactive metals from aqueous mixtures. After the radioisotopes are absorbed onto the ferric hydroxide, the resulting sludge can be placed in a metal drum before being mixed with cement to form a solid waste form. In order to get better long-term performance (mechanical stability) from such forms, they may be made from a mixture of fly ash, or blast furnace slag, and Portland cement, instead of normal concrete (made with Portland cement, gravel and sand).

Synroc

The Australian Synroc (synthetic rock) is a more sophisticated way to immobilize such waste, and this process may eventually come into commercial use for civil wastes (it is currently being developed for US military wastes). Synroc was invented by Prof Ted Ringwood (a geochemist) at the Australian National University. The Synroc contains pyrochlore and cryptomelane type minerals. The original form of Synroc (Synroc C) was designed for the liquid high level waste (PUREX raffinate) from a light water reactor. The main minerals in this Synroc are hollandite ($BaAl_2Ti_6O_{16}$), zirconolite ($CaZrTi_2O_7$) and perovskite ($CaTiO_3$). The zirconolite and perovskite are hosts for the actinides. The strontium and barium will be fixed in the perovskite. The caesium will be fixed in the hollandite.

Long Term Management

The time frame in question when dealing with radioactive waste ranges from 10,000 to 1,000,000 years, according to studies based on the effect of estimated radiation doses. Researchers suggest that forecasts of health detriment for such periods should be examined critically. Practical studies only consider up to 100 years as far as effective planning and cost evaluations are concerned. Long term behavior of radioactive wastes remains a subject for ongoing research projects in geoforecasting.

Above-ground Disposal

Dry cask storage typically involves taking waste from a spent fuel pool and sealing it (along with an inert gas) in a steel cylinder, which is placed in a concrete cylinder which acts as a radiation shield. It is a relatively inexpensive method which can be done at a central facility or adjacent to the source reactor. The waste can be easily retrieved for reprocessing.

Geologic Disposal

The process of selecting appropriate deep final repositories for high level waste and spent fuel is now under way in several countries with the first expected to be commissioned some time after 2010. The basic concept is to locate a large, stable geologic formation and use mining technology to excavate a tunnel, or large-bore tunnel boring machines (similar to those used to drill the Channel Tunnel from England to France) to drill a shaft 500 metres (1,600 ft) to 1,000 metres (3,300 ft) below the surface where rooms or vaults can be excavated for disposal of high-level radioactive waste. The goal is to permanently isolate nuclear waste from the human environment.

Many people remain uncomfortable with the immediate stewardship cessation of this disposal system, suggesting perpetual management and monitoring would be more prudent.

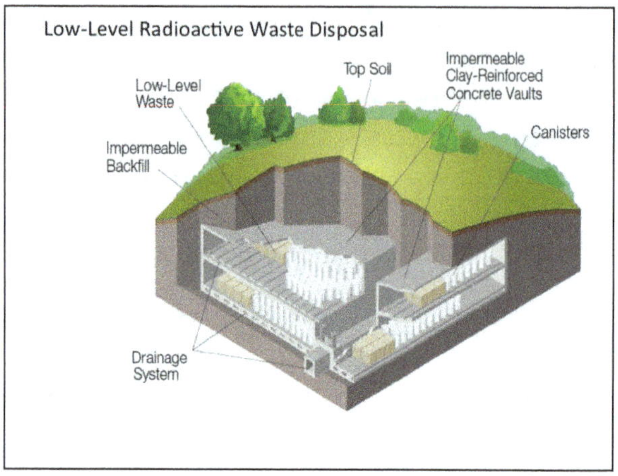

Diagram of an underground low-level radioactive waste disposal site.

Radioactive materials at the leaked from a damaged storage drum.

On Feb. 14, 2014, radioactive materials at the Waste Isolation Pilot Plant leaked from a damaged storage drum due to the use of incorrect packing material. Analysis showed the lack of a "safety culture" at the plant since its successful operation for 15 years had bred complacency.

Because some radioactive species have half-lives longer than one million years, even very low container leakage and radionuclide migration rates must be taken into account. Moreover, it may require more than one half-life until some nuclear materials lose enough radioactivity to cease being lethal to living things. A 1983 review of the Swedish radioactive waste disposal program by the National Academy of Sciences found that country's estimate of several hundred thousand years—perhaps up to one million years—being necessary for waste isolation "fully justified."

Ocean floor disposal of radioactive waste has been suggested by the finding that deep waters in the North Atlantic Ocean do not present an exchange with shallow waters for about 140 years based on oxygen content data recorded over a period of 25 years. They include burial beneath a stable abyssal plain, burial in a subduction zone that would slowly carry the waste downward into the Earth's mantle, and burial beneath a remote natural or human-made island. While these approaches all have merit and would facilitate an international solution to the problem of disposal of radioactive waste, they would require an amendment of the Law of the Sea.

The proposed land-based subductive waste disposal method disposes of nuclear waste in a subduction zone accessed from land and therefore is not prohibited by international agreement. This method has been described as the most viable means of disposing of radioactive waste, and as the state-of-the-art as of 2001 in nuclear waste disposal technology. Another approach termed Remix & Return would blend high-level waste with uranium mine and mill tailings down to the level of the original radioactivity of the uranium ore, then replace it in inactive uranium mines. This approach has the merits of providing jobs for miners who would double as disposal staff, and of facilitating a cradle-to-grave cycle for radioactive materials, but would be inappropriate for spent reactor fuel in the absence of reprocessing, due to the presence of highly toxic radioactive elements such as plutonium within it.

Deep borehole disposal is the concept of disposing of high-level radioactive waste from nuclear reactors in extremely deep boreholes. Deep borehole disposal seeks to place the waste as much as 5 kilometres (3.1 mi) beneath the surface of the Earth and relies primarily on the immense natural geological barrier to confine the waste safely and permanently so that it should never pose a threat to the environment. The Earth's crust contains 120 trillion tons of thorium and 40 trillion tons of uranium (primarily at relatively trace concentrations of parts per million each adding up over the crust's 3×10^{19} ton mass), among other natural radioisotopes. Since the fraction of nuclides decaying per unit of time is inversely proportional to an isotope's half-life, the relative radioactivity of the lesser amount of human-produced radioisotopes (thousands of tons instead of trillions of tons) would diminish once the isotopes with far shorter half-lives than the bulk of natural radioisotopes decayed.

In January 2013, Cumbria county council rejected UK central government proposals to start work on an underground storage dump for nuclear waste near to the Lake District National Park. "For any host community, there will be a substantial community benefits package and worth hundreds of millions of pounds" said Ed Davey, Energy Secretary, but nonetheless, the local elected body voted 7–3 against research continuing, after hearing evidence from independent geologists that "the fractured strata of the county was impossible to entrust with such dangerous material and a hazard lasting millennia."

Transmutation

There have been proposals for reactors that consume nuclear waste and transmute it to other, less-harmful or shorter-lived, nuclear waste. In particular, the Integral Fast Reactor was a proposed nuclear reactor with a nuclear fuel cycle that produced no transuranic waste and in fact, could consume transuranic waste. It proceeded as far as large-scale tests, but was then canceled by the US Government. Another approach, considered safer but requiring more development, is to dedicate subcritical reactors to the transmutation of the left-over transuranic elements.

An isotope that is found in nuclear waste and that represents a concern in terms of proliferation is Pu-239. The large stock of plutonium is a result of its production inside uranium-fueled reactors and of the reprocessing of weapons-grade plutonium during the weapons program. An option for getting rid of this plutonium is to use it as a fuel in a traditional Light Water Reactor (LWR). Several fuel types with differing plutonium destruction efficiencies are under study.

Transmutation was banned in the US in April 1977 by President Carter due to the danger of plutonium proliferation, but President Reagan rescinded the ban in 1981. Due to the economic losses and risks, construction of reprocessing plants during this time did not resume. Due to high energy demand, work on the method has continued in the EU. This has resulted in a practical nuclear research reactor called Myrrha in which transmutation is possible. Additionally, a new research program called ACTINET has been started in the EU to make transmutation possible on a large, industrial scale. According to President Bush's Global Nuclear Energy Partnership (GNEP) of 2007, the US is now actively promoting research on transmutation technologies needed to markedly reduce the problem of nuclear waste treatment.

There have also been theoretical studies involving the use of fusion reactors as so called "actinide burners" where a fusion reactor plasma such as in a tokamak, could be "doped" with a small amount of the "minor" transuranic atoms which would be transmuted (meaning fissioned in the actinide case) to lighter elements upon their successive bombardment by the very high energy neutrons produced by the fusion of deuterium and tritium in the reactor. A study at MIT found that only 2 or 3 fusion reactors with parameters similar to that of the International Thermonuclear Experimental Reactor (ITER) could transmute the entire annual minor actinide production from all of the light water reactors presently operating in the United States fleet while simultaneously generating approximately 1 gigawatt of power from each reactor.

Re-use

Another option is to find applications for the isotopes in nuclear waste so as to re-use them. Already, caesium-137, strontium-90 and a few other isotopes are extracted for certain industrial applications such as food irradiation and radioisotope thermoelectric generators. While re-use does not eliminate the need to manage radioisotopes, it can reduce the quantity of waste produced.

The Nuclear Assisted Hydrocarbon Production Method, Canadian patent application 2,659,302, is a method for the temporary or permanent storage of nuclear waste materials comprising the placing of waste materials into one or more repositories or boreholes constructed into an unconventional oil formation. The thermal flux of the waste materials fracture the formation and alters the chemical and physical properties of hydrocarbon material within the subterranean formation to allow removal of the altered material. A mixture of hydrocarbons, hydrogen, and other formation fluids is produced from the formation. The radioactivity of high-level radioactive waste affords proliferation resistance to plutonium placed in the periphery of the repository or the deepest portion of a borehole.

Breeder reactors can run on U-238 and transuranic elements, which comprise the majority of spent fuel radioactivity in the 1,000–100,000-year time span.

Space Disposal

Space disposal is attractive because it removes nuclear waste from the planet. It has significant disadvantages, such as the potential for catastrophic failure of a launch vehicle, which could spread radioactive material into the atmosphere and around the world. A high number of launches would be required because no individual rocket would be able to carry very much of the material relative to the total amount that needs to be disposed of. This makes the proposal impractical economically and it increases the risk of at least one or more launch failures. To further complicate matters, international agreements on the regulation of such a program would need to be established. Costs and inadequate reliability of modern rocket launch systems for space disposal has been one of the motives for interest in non-rocket space launch systems such as mass drivers, space elevators, and other proposals.

National Management Plans

Most countries are considerably ahead of the United States in developing plans for high-level radioactive waste disposal. Sweden and Finland are furthest along in committing to a particular disposal technology, while many others reprocess spent fuel or contract with France or Great Britain to do it, taking back the resulting plutonium and high-level waste. "An increasing backlog of plutonium from reprocessing is developing in many countries. It is doubtful that reprocessing makes economic sense in the present environment of cheap uranium."

In many European countries (e.g., Britain, Finland, the Netherlands, Sweden and Switzerland) the risk or dose limit for a member of the public exposed to radiation from a future high-level nuclear waste facility is considerably more stringent than that suggested by the International Commission on Radiation Protection or proposed in the United States. European limits are often more stringent than the standard suggested in 1990 by the International Commission on Radiation Protection by a factor of 20, and more stringent by a factor of ten than the standard proposed by the US Environmental Protection Agency (EPA) for Yucca Mountain nuclear waste repository for the first 10,000 years after closure.

The U.S. EPA's proposed standard for greater than 10,000 years is 250 times more permissive than the European limit. The U.S. EPA proposed a legal limit of a maximum of 3.5 millisieverts (350 millirem) each annually to local individuals after 10,000 years, which would be up to several percent of the exposure currently received by some populations in the highest natural background regions on Earth, though the U.S. DOE predicted that received dose would be much below that limit. Over a timeframe of thousands of years, after the most active short half-life radioisotopes decayed, burying U.S. nuclear waste would increase the radioactivity in the top 2000 feet of rock and soil in the United States (10 million km^2) by $\approx$ 1 part in 10 million over the cumulative amount of natural radioisotopes in such a volume, but the vicinity of the site would have a far higher concentration of artificial radioisotopes underground than such an average.

Mongolia

After serious opposition had arisen about plans and negotiations between Mongolia with Japan and the United States of America to build nuclear-waste facilities in Mongolia, Mongolia stopped all negotiations in September 2011. These negotiations had started after U.S. Deputy Secretary of Energy Daniel B. Poneman visited Mongolia in September, 2010. Talks took place in Washington DC between officials of Japan, the United States and Mongolia in February 2011. After this the United Arab Emirates (UAE), which wanted to buy nuclear fuel from Mongolia, joined in the negotiations. The talks were kept secret, and although The Mainichi Daily News reported on them in May, Mongolia officially denied the existence of these negotiations. However, alarmed by this news, Mongolian citizens protested against the plans, and demanded the government withdraw the plans and disclose information. The Mongolian President Tsakhiagiin Elbegdorj issued a presidential order on September 13, banning all negotiations with foreign governments or international organizations on nuclear-waste storage plans in Mongolia. The Mongolian government has accused the newspaper of distributing false claims around the world. After the presidential order, the Mongolian president fired the individual who was supposedly involved in these conversations.

Illegal Dumping

Authorities in Italy are investigating a 'Ndrangheta mafia clan accused of trafficking and illegally dumping nuclear waste. According to a whistleblower, a manager of the Italy's state energy research agency Enea paid the clan to get rid of 600 drums of toxic and radioactive waste from Italy,

Switzerland, France, Germany, and the US, with Somalia as the destination, where the waste was buried after buying off local politicians. Former employees of Enea are suspected of paying the criminals to take waste off their hands in the 1980s and 1990s. Shipments to Somalia continued into the 1990s, while the Ndrangheta clan also blew up shiploads of waste, including radioactive hospital waste, sending them to the sea bed off the Calabrian coast. According to the environmental group Legambiente, former members of the Ndrangheta have said that they were paid to sink ships with radioactive material for the last 20 years.

Accidents

A few incidents have occurred when radioactive material was disposed of improperly, shielding during transport was defective, or when it was simply abandoned or even stolen from a waste store. In the Soviet Union, waste stored in Lake Karachay was blown over the area during a dust storm after the lake had partly dried out. At Maxey Flat, a low-level radioactive waste facility located in Kentucky, containment trenches covered with dirt, instead of steel or cement, collapsed under heavy rainfall into the trenches and filled with water. The water that invaded the trenches became radioactive and had to be disposed of at the Maxey Flat facility itself. In other cases of radioactive waste accidents, lakes or ponds with radioactive waste accidentally overflowed into the rivers during exceptional storms. In Italy, several radioactive waste deposits let material flow into river water, thus contaminating water for domestic use. In France, in the summer of 2008 numerous incidents happened; in one, at the Areva plant in Tricastin, it was reported that during a draining operation, liquid containing untreated uranium overflowed out of a faulty tank and about 75 kg of the radioactive material seeped into the ground and, from there, into two rivers nearby; in another case, over 100 staff were contaminated with low doses of radiation.

Scavenging of abandoned radioactive material has been the cause of several other cases of radiation exposure, mostly in developing nations, which may have less regulation of dangerous substances (and sometimes less general education about radioactivity and its hazards) and a market for scavenged goods and scrap metal. The scavengers and those who buy the material are almost always unaware that the material is radioactive and it is selected for its aesthetics or scrap value. Irresponsibility on the part of the radioactive material's owners, usually a hospital, university or military, and the absence of regulation concerning radioactive waste, or a lack of enforcement of such regulations, have been significant factors in radiation exposures.

Transportation accidents involving spent nuclear fuel from power plants are unlikely to have serious consequences due to the strength of the spent nuclear fuel shipping casks.

On 15 December 2011, top government spokesman Osamu Fujimura of the Japanese government admitted that nuclear substances were found in the waste of Japanese nuclear facilities. Although Japan did commit itself in 1977 to these inspections in the safeguard agreement with the IAEA, the reports were kept secret for the inspectors of the International Atomic Energy Agency. Japan did start discussions with the IAEA about the large quantities of enriched uranium and plutonium that were discovered in nuclear waste cleared away by Japanese nuclear operators. At the press conference Fujimura said: "Based on investigations so far, most nuclear substances have been properly managed as waste, and from that perspective, there is no problem in safety management," But according to him, the matter was at that moment still being investigated.

References

- Williams, i.d.; kelly, j. (2003). "green waste collection and the public's recycling behaviour in the borough of wyre, england". Resources, conservation and recycling. 38 (2): 139–159. Doi:10.1016/s0921-3449(02)00106-4. Issn 0921-3449
- What-is-waste, waste-management, environment-society: enviroliteracy.org, Retrieved 5 January, 2019
- "Sewerage – definition of sewerage in english from the oxford dictionary". Oxforddictionaries.com. Archived from the original on 2015-09-24. Retrieved 2015-09-04
- Biodegradable-waste: appropedia.org, Retrieved 6 February, 2019
- Bellemare, marc f.; çakir, metin; peterson, hikaru hanawa; novak, lindsey; rudi, jeta (2017). "on the measurement of food waste". American journal of agricultural economics. 99 (5): 1148–1158. Doi:10.1093/ajae/aax034
- Industrial-waste-types, chemistry: byjus.com, Retrieved 7 March, 2019

2

Types of Waste Management

Waste management includes the activities and methods used to manage different types of waste. Solid waste management, agricultural waste management, slaughterhouse waste management and organic waste management are the various types of waste management. The chapter closely examines these types of waste management to provide an extensive understanding of the subject.

WASTE MANAGEMENT

Waste management is the collection, transport, processing or disposal of waste materials, usually ones produced by human activity, in an effort to reduce their effect on human health or local aesthetics or amenity.

Techniques of Waste Management

Following are the effective methods of waste management except ocean dumping which has been controlled, regulated and banned in some cases in order to stop the hazardous materials to be dumped into the sea:

- Ocean Dumping.
- Sanitary Landfill.
- Incineration.
- Composting.
- Waste separation, recycling and recovery.
- Mechanical and biological waste treatment.
- Mechanical sorting of wastes.

Importance of Waste Management

According to some research reports, every year, 62 million tonnes of waste is generating out of which only 28% of waste is recycled and 72% is left on the roads and the landfill areas for several years, which is fraught with hazardous consequences. Presently, most of the developing countries are facing almost similar situation. Therefore, segregating various kinds of wastes from the households, efficient waste collection systems, proper disposal, and sustainable recovery are very much needed processes in modern-day world.

Nothing is waste until it cannot be used anymore in any way. We generally see wastes piled up but do not think deep enough on how to use them. Today there are many innovations available in waste management like recycling wastes into usable products, generating methane or fuels, manufacturing new products for home/commercial usage such as fence posts, furniture and so on. Therefore, the importance of managing waste in a very effective way has enhanced many folds now-a-days.

Manage Waste in a Scientific Way

An effective strategy to dispose various wastes can offer many solutions for the various problems associated with waste materials. The most efficient way to deal the waste is to recycle and compost. It ensures there is gradual improvement of new and cost-effective facilities, which aim to encourage higher environmental protection standards.

Efficient Management of Landfills: Most landfills lack proper on-site waste management thus contributing to some additional threats to the environment. After certain period, these landfills tend to leak and pollute ground water and other neighbouring environmental habitats making waste management a very difficult task. These landfills also generate potentially unsafe gases.

Most of the laws and regulation guiding the operations of landfills are often slack at monitoring and regulating the different types of wastes like medical waste, municipal waste, special waste, or hazardous waste. This slackness of the laws contributes towards significant increases in toxicity and hazardous waste to a point where the landfill wastes problems often lasts up to many decades.

Concept of 3R: Keeping the nature of waste generated by the households and industries in view, the three R's mantra of Reuse, Reduce and Recycle appears be more suitable to meet this challenge. Care should be taken to see that these options should be taken up under a well sustainable framework.

Zero Waste System: Industries can play a very significant role by streamlining the process of waste management for the entire organization through the one central system. Companies can deal effectively with hazardous waste regulations about the generation of their waste, its handling, transportation and safe disposal of the waste. In fact they can very well attain "zero waste to landfill" targets as part of sustainability initiatives.

More emphasis should be given to responsible resource use with an objective of avoidance, maximizing recycling and waste reduction methods. This involves techniques such as repair of broken things instead of buying new, purchasing and re-using second-hand items, and designing reusable and recyclable products.

Benefits of Waste Management

The benefits of managing wastes are very clear from the following:

Saving Natural Resources: It is a matter of serious concern for us when we see that many natural resources such as trees, gas, and water are diminishing very rapidly. We all know that paper, cupboards, paper cups, and many other products are made from trees. However, trees are cut down on a large scale every year and new trees are not replacing them at the rate they are cut. Therefore, we should think to recycle paper products so that there is no need of cutting new trees. It is also quite possible to reuse plastic and metal items. Some countries have established certain sites where people can bring and sell old newspapers, metal items, or glass. These products are recycled and used again.

Producing energy: Recycling is a great way to produce energy. By recycling something, we save energy because more energy is usually needed to produce a new item. For example, it is possible to get energy from recycling the waste. In Mukesh Ambani's house, garbage is used to create electricity. It is reported that electricity is generating from the waste by a special system in their house. First, the dry and wet wastes are separated, after which the electricity is generated. Electricity in such a large house is being produced by recycling the wastes.

Reduce pollution: Recycling is a one of the most powerful tools to save nature and humanity. Making more people aware is a contribution in a better future. The more people start to manage their waste, the better planet we will have to live in. Apart from leaving a lot of waste, humanity pollutes the environment by producing various products. Factories considerably pollute the atmosphere with smoke by manufacturing processes. Recycling reduces pollution and helps save energy. Sounds like a win-win solution.

Recycling Rubbish: Rubbish is a big problem for aquatic life too. A lot of rubbish is thrown away into the sea and ocean. There are large areas of waste called "waste islands," which are made of rubbish that has been accumulated in one place. Recycling rubbish is important for nature and humanity. Recycling begins with rubbish management. It means that paper items can be collected into a paper bin for paper waste, glass into a glass bin and so on.

Indian Scenario on Waste Management

Under Indian scenario, the approach towards solid waste management is not scientific, at least for a common person. As the number of industries is increasing day by day due to a rapid increase in urbanization, this is eventually leading to the generation of municipal solid waste at a significantly high rate. In today's time, this problem has become so disturbing that a small rain creates conditions like floods in towns and cities.

If we look at the condition that was existing during few decades ago, it gives us an impression that it was far better than the present. During those days not only, the roads but even the streets of small towns used to be cleaned every day. Cleaning of drainages used to a part of everyday life and generally, no one would deliberately put garbage in the drainage. The silts and litter collected from the houses used to convert into organic fertilizers.

In this context, we can cite one example of a small city of Gujarat state–Bhavnagar to have a very good system of managing the drains. Bhavnagar is one of the few cities having underground drainage facilities ever since the time prior to independence. The old city of Bhavnagar had underground drainage system. These internal drainage works of the city came into practice since 1936 by Public Works Department of the state.

During olden days, people had a habit of shopping with a cloth bag. The goods purchased from the market used to be brought home usually in this bag. However, gradually the bag was replaced by a plastic bag. People started feeling ashamed to walk around with a cloth bag, as the use of plastic bags became the national habit. Now the use of plastic is growing due to continuous consumption of various commodities of day-to-day life.

Solutions for Waste Management

We are moving from worse to worst in waste management because of the lack of laws and regulations in this regard. We do not have even a single city where the entire waste of the city is being managed in a scientific way. Identifying the sites where waste could be dumped and managed is becoming a great problem because of the resistance of the local people due to environmental considerations. In most of the towns, this waste ultimately finds its way on the banks of the rivers or on the periphery of water bodies. As a result, the surface water bodies are getting polluted. It is imperative that city planners consider this aspect in the planning stage itself.

Adequate consideration should be given to monitor the processes. There is a need to educate the public and plan in a proper way for waste management programs and implement into current waste management system. Some awareness programs also must be conducted to enhance the knowledge about managing the waste in an eco-friendly way. If these things are accomplished properly, we will be able to save money at individual as well as society level.

The concerned authorities should provide the introduction of certain programs and policy development about how efficiently we can do the waste management. It is not that everything eventually will decompose as this process results in many non-worthy by-products in the end. Most of the things we discard do not decompose. We unwittingly contribute to store them in a huge quantity at underground level. The time demands that waste management must be sustainable. Now a days, the proper management of solid wastes needs an economically affordable, socially accepted and environmentally friendly appropriate technology.

Recycling makes sense in most of the areas of our economy. In modern cities, wastewater is treated to use it again but when cooking oil starts solidifying in collection systems and eventually hamper the flow of wastewater, we have to deal with very high costs for repairs. Moreover, not all kinds of wastes are getting recycled. Only metals are recycled almost 100% into consumer goods. Glass and paper are at the second place but unfortunately, only less than 50% of plastics are being recycled. A very fruitful alternative to this problem could be to use bio-plastic, which tend to be more favourable to the environment.

Various communities from all walks of life whether urban or rural should be educated about the harmful impact of waste on infrastructure and public health through multimedia and how waste collection services can be improved to at local level. People should understand if they are not able to manage them, these waste materials might cause a serious hazard to human health as well as to the environment.

In fact, it is rarely better to simply throw something away than to reuse it or recycle it. We must always have in mind that our main and ultimate objective of waste management program is to keep the Earth clean and to conduct our activities in an eco-friendly way so that we can leave a more hygienic and healthy environment for future generations.

SOLID WASTE MANAGEMENT

Solid waste management is the collecting, treating, and disposing of solid material that is discarded because it has served its purpose or is no longer useful. Improper disposal of municipal solid waste can create unsanitary conditions, and these conditions in turn can lead to pollution of the environment and to outbreaks of vector-borne disease—that is, diseases spread by rodents and insects. The tasks of solid-waste management present complex technical challenges. They also pose a wide variety of administrative, economic, and social problems that must be managed and solved.

Developments in Waste Management

A technological approach to solid-waste management began to develop in the latter part of the 19th century. Watertight garbage cans were first introduced in the United States, and sturdier vehicles were used to collect and transport wastes. A significant development in solid-waste treatment and disposal practices was marked by the construction of the first refuse incinerator in England in 1874. By the beginning of the 20th century, 15 percent of major American cities were incinerating solid waste. Even then, however, most of the largest cities were still using primitive disposal methods such as open dumping on land or in water.

Technological advances continued during the first half of the 20th century, including the development of garbage grinders, compaction trucks, and pneumatic collection systems. By mid-century, however, it had become evident that open dumping and improper incineration of solid waste were causing problems of pollution and jeopardizing public health. As a result, sanitary landfills were developed to replace the practice of open dumping and to reduce the reliance on waste incineration. In many countries waste was divided into two categories: hazardous and nonhazardous, and separate regulations were developed for their disposal. Landfills were designed and operated in a manner that minimized risks to public health and the environment. New refuse incinerators were designed to recover heat energy from the waste and were provided with extensive air pollution control devices to satisfy stringent standards of air quality. Modern solid-waste management plants in most developed countries now emphasize the practice of recycling and waste reduction at the source rather than incineration and land disposal.

Solid-waste Characteristics

Composition and Properties

The sources of solid waste include residential, commercial, institutional, and industrial activities. Certain types of wastes that cause immediate danger to exposed individuals or environments are classified as hazardous; these are discussed in the article hazardous-waste management. All nonhazardous solid waste from a community that requires collection and transport to a processing or disposal site is called refuse or municipal solid waste (MSW). Refuse includes garbage and rubbish. Garbage is mostly decomposable food waste; rubbish is mostly dry material such as glass, paper, cloth, or wood. Garbage is highly putrescible or decomposable, whereas rubbish is not. Trash is rubbish that includes bulky items such as old refrigerators, couches, or large tree stumps. Trash requires special collection and handling.

Construction and demolition (C&D) waste (or debris) is a significant component of total solid waste quantities (about 20 percent in the United States), although it is not considered to be part of the MSW stream. However, because C&D waste is inert and nonhazardous, it is usually disposed of in municipal sanitary landfills.

Another type of solid waste, perhaps the fastest-growing component in many developed countries, is electronic waste, or e-waste, which includes discarded computer equipment, televisions, telephones, and a variety of other electronic devices. In 2006 e-waste made up 5 percent of the total solid waste stream, and the United Nations Environment Programme estimated that developed countries would triple their output of e-waste by 2010. Concern over this type of waste is escalating. Lead, mercury, and cadmium are among the materials of concern in electronic devices, and governmental policies may be required to regulate their recycling and disposal.

Solid-waste characteristics vary considerably among communities and nations. American refuse is usually lighter, for example, than European or Japanese refuse. In the United States paper and paperboard products make up close to 40 percent of the total weight of MSW; food waste accounts for less than 10 percent. The rest is a mixture of yard trimmings, wood, glass, metal, plastic, leather, cloth, and other miscellaneous materials. In a loose or uncompacted state, MSW of this type weighs approximately 120 kg per cubic metre (200 pounds per cubic yard). These figures vary with geographic location, economic conditions, season of the year, and many other factors. Waste characteristics from each community must be studied carefully before any treatment or disposal facility is designed and built.

Generation and Storage

Rates of solid-waste generation vary widely. In the United States, for example, municipal refuse is generated at an average rate of approximately 2 kg (4.4 pounds) per person per day. Japan generates roughly half this amount, yet in Canada the rate is 3 kg (almost 7 pounds) per person per day. In some developing countries (e.g. India) the average rate can be lower than 0.5 kg (1 pound) per person per day. These data include refuse from commercial, institutional, and industrial as well as residential sources. The actual rates of refuse generation must be carefully determined when a community plans a solid-waste management project.

Most communities require household refuse to be stored in durable, easily cleaned containers with tight-fitting covers in order to minimize rodent or insect infestation and offensive odours. Galvanized metal or plastic containers of about 115-litre (30-gallon) capacity are commonly used, although some communities employ larger containers that can be mechanically lifted and emptied into collection trucks. Plastic bags are frequently used as liners or as disposable containers for curbside collection. Where large quantities of refuse are generated—such as at shopping centres, hotels, or apartment buildings—dumpsters may be used for temporary storage until the waste is collected. Some office and commercial buildings use on-site compactors to reduce the waste volume.

Solid-waste Collection

Collecting and Transporting

Proper solid-waste collection is important for the protection of public health, safety, and environmental quality. It is a labour-intensive activity, accounting for approximately three-quarters of the total cost of solid-waste management. Public employees are often assigned to the task, but sometimes it is more economical for private companies to do the work under contract to the municipality or for private collectors to be paid by individual home owners. A driver and one or two loaders serve each collection vehicle. These are typically trucks of the enclosed, compacting type, with capacities up to 30 cubic metres (40 cubic yards). Loading can be done from the front, rear, or side. Compaction reduces the volume of refuse in the truck to less than half of its loose volume.

The task of selecting an optimal collection route is a complex problem, especially for large and densely populated cities. An optimal route is one that results in the most efficient use of labour and equipment, and selecting such a route requires the application of computer analyses that account for all the many design variables in a large and complex network. Variables include frequency of collection, haulage distance, type of service, and climate. Collection of refuse in rural areas can present a special problem, since the population densities are low, leading to high unit costs.

Refuse collection usually occurs at least once per week because of the rapid decomposition of food waste. The amount of garbage in the refuse of an individual home can be reduced by garbage grinders, or garbage disposals. Ground garbage puts an extra load on sewerage systems, but this can usually be accommodated. Many communities now conduct source separation and recycling programs, in which homeowners and businesses separate recyclable materials from garbage and place them in separate containers for collection. In addition, some communities have drop-off centres where residents can bring recyclables.

Transfer Stations

If the final destination of the refuse is not near the community in which it is generated, one or more transfer stations may be necessary. A transfer station is a central facility where refuse from many collection vehicles is combined into a larger vehicle, such as a tractor-trailer unit. Open-top trailers are designed to carry about 76 cubic metres (100 cubic yards) of uncompacted waste to a regional processing or disposal location. Closed compactor-type trailers are also available, but they must be equipped with ejector mechanisms. In a direct discharge type of station, several collection trucks empty directly into the transport vehicle. In a storage discharge type of station, refuse is first emptied into a storage pit or onto a platform, and then machinery is used to hoist or push the solid waste into the transport vehicle. Large transfer stations can handle more than 500 tons of refuse per day.

Solid-waste Treatment and Disposal

Once collected, municipal solid waste may be treated in order to reduce the total volume and weight of material that requires final disposal. Treatment changes the form of the waste and makes it easier to handle. It can also serve to recover certain materials, as well as heat energy, for recycling or reuse.

Incineration

Furnace Operation

Burning is a very effective method of reducing the volume and weight of solid waste. In modern incinerators the waste is burned inside a properly designed furnace under very carefully controlled conditions. The combustible portion of the waste combines with oxygen, releasing mostly carbon dioxide, water vapour, and heat. Incineration can reduce the volume of uncompacted waste by more than 90 percent, leaving an inert residue of ash, glass, metal, and other solid materials called bottom ash. The gaseous by-products of incomplete combustion, along with finely divided particulate material called fly ash, are carried along in the incinerator airstream. Fly ash includes cinders, dust, and soot. In order to remove fly ash and gaseous by-products before they are exhausted into the atmosphere, modern incinerators must be equipped with extensive emission control devices. Such devices include fabric baghouse filters, acid gas scrubbers, and electrostatic precipitators. Bottom ash and fly ash are usually combined and disposed of in a landfill. If the ash is found to contain toxic metals, it must be managed as a hazardous waste.

Municipal solid-waste incinerators are designed to receive and burn a continuous supply of refuse. A deep refuse storage pit, or tipping area, provides enough space for about one day of waste storage. The refuse is lifted from the pit by a crane equipped with a bucket or grapple device. It is then deposited into a hopper and chute above the furnace and released onto a charging grate or stoker. The grate shakes and moves waste through the furnace, allowing air to circulate around the burning material. Modern incinerators are usually built with a rectangular furnace, although rotary kiln furnaces and vertical circular furnaces are available. Furnaces are constructed of refractory bricks that can withstand the high combustion temperatures.

Combustion in a furnace occurs in two stages: primary and secondary. In primary combustion, moisture is driven off, and the waste is ignited and volatilized. In secondary combustion, the remaining unburned gases and particulates are oxidized, eliminating odours and reducing the amount of fly ash in the exhaust. When the refuse is very moist, auxiliary gas or fuel oil is sometimes burned to start the primary combustion.

In order to provide enough oxygen for both primary and secondary combustion, air must be thoroughly mixed with the burning refuse. Air is supplied from openings beneath the grates or is admitted to the area above. The relative amounts of this underfire air and overfire air must be determined by the plant operator to achieve good combustion efficiency. A continuous flow of air can be maintained by a natural draft in a tall chimney or by mechanical forced-draft fans.

Energy Recovery

The energy value of refuse can be as much as one-third that of coal, depending on the paper content, and the heat given off during incineration can be recovered by the use of a refractory-lined furnace coupled to a boiler. Boilers convert the heat of combustion into steam or hot water, thus allowing the energy content of the refuse to be recycled. Incinerators that recycle heat energy in this way are called waste-to-energy plants. Instead of a separate furnace and boiler, a water-tube wall furnace may also be used for energy recovery. Such a furnace is lined with vertical steel tubes spaced closely enough to form continuous sections of wall. The walls are insulated on the outside in order to reduce heat loss. Water circulating through the tubes absorbs heat to produce steam, and it also helps to control combustion temperatures without the need for excessive air, thus lowering air pollution control costs.

Waste-to-energy plants operate as either mass burn or refuse-derived fuel systems. A mass burn system uses all the refuse, without prior treatment or preparation. A refuse-derived fuel system separates combustible wastes from noncombustibles such as glass and metal before burning. If a turbine is installed at the plant, both steam and electricity can be produced in a process called cogeneration.

Waste-to-energy systems are more expensive to build and operate than plain incinerators because of the need for special equipment and controls, highly skilled technical personnel, and auxiliary fuel systems. On the other hand, the sale of generated steam or electricity offsets much of the extra cost, and recovery of heat energy from refuse is a viable solid-waste management option from both an engineering and an economic point of view. About 80 percent of municipal refuse incinerators in the United States are waste-to-energy facilities.

Composting

Another method of treating municipal solid waste is composting, a biological process in which the organic portion of refuse is allowed to decompose under carefully controlled conditions. Microbes metabolize the organic waste material and reduce its volume by as much as 50 percent. The stabilized product is called compost or humus. It resembles potting soil in texture and odour and may be used as a soil conditioner or mulch.

Composting offers a method of processing and recycling both garbage and sewage sludge in one operation. As more stringent environmental rules and siting constraints limit the use of solid-waste incineration and landfill options, the application of composting is likely to increase. The steps involved in the process include sorting and separating, size reduction, and digestion of the refuse.

Sorting and Shredding

The decomposable materials in refuse are isolated from glass, metal, and other inorganic items through sorting and separating operations. These are carried out mechanically, using differences in such physical characteristics of the refuse as size, density, and magnetic properties. Shredding or pulverizing reduces the size of the waste articles, resulting in a uniform mass of material. It is accomplished with hammer mills and rotary shredders.

Digesting and Processing

Pulverized waste is ready for composting either by the open windrow method or in an enclosed mechanical facility. Windrows are long, low mounds of refuse. They are turned or mixed every few days to provide air for the microbes digesting the organics. Depending on moisture conditions, it may take five to eight weeks for complete digestion of the waste. Because of the metabolic action of aerobic bacteria, temperatures in an active compost pile reach about 65 °C (150 °F), killing pathogenic organisms that may be in the waste material.

Open windrow composting requires relatively large land areas. Enclosed mechanical composting facilities can reduce land requirements by about 85 percent. Mechanical composting systems employ one or more closed tanks or digesters equipped with rotating vanes that mix and aerate the shredded waste. Complete digestion of the waste takes about one week.

Digested compost must be processed before it can be used as a mulch or soil conditioner. Processing includes drying, screening, and granulating or pelletizing. These steps improve the market value of the compost, which is the most serious constraint to the success of composting as a waste management option. Agricultural demand for digested compost is usually low because of the high cost of transporting it and because of competition with inorganic chemical fertilizers.

Sanitary Landfill

Land disposal is the most common management strategy for municipal solid waste. Refuse can be safely deposited in a sanitary landfill, a disposal site that is carefully selected, designed, constructed, and operated to protect the environment and public health. One of the most important factors relating to landfilling is that the buried waste never comes in contact with surface water or groundwater. Engineering design requirements include a minimum distance between the bottom of the landfill and the seasonally high groundwater table. Most new landfills are required to have an impermeable liner or barrier at the bottom, as well as a system of groundwater-monitoring wells. Completed landfill sections must be capped with an impermeable cover to keep precipitation or surface runoff away from the buried waste. Bottom and cap liners may be made of flexible plastic membranes, layers of clay soil, or a combination of both.

Constructing the Landfill

The basic element of a sanitary landfill is the refuse cell. This is a confined portion of the site in which refuse is spread and compacted in thin layers. Several layers may be compacted on top of one another to a maximum depth of about 3 metres (10 feet). The compacted refuse occupies about one-quarter of its original loose volume. At the end of each day's operation, the refuse is covered with a layer of soil to eliminate windblown litter, odours, and insect or rodent problems. One refuse cell thus contains the daily volume of compacted refuse and soil cover. Several adjacent refuse cells make up a lift, and eventually a landfill may comprise two or more lifts stacked one on top of the other. The final cap for a completed landfill may also be covered with a layer of topsoil that can support vegetative growth.

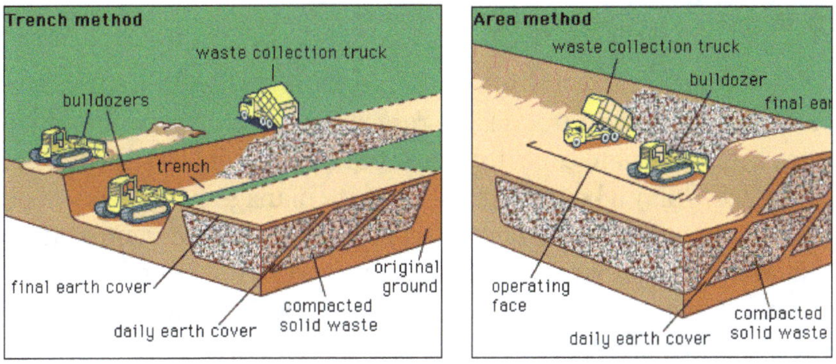

Two methods of constructing a sanitary landfill.

Daily cover soil may be available on-site, or it may be hauled in and stockpiled from off-site sources. Various types of heavy machinery, such as crawler tractors or rubber-tired dozers, are used to spread and compact the refuse and soil. Heavy steel-wheeled compactors may also be employed to achieve high-density compaction of the refuse.

The area and depth of a new landfill are carefully staked out, and the base is prepared for construction of any required liner and leachate-collection system. Where a plastic liner is used, at least 30 cm (12 inches) of sand is carefully spread over it to provide protection from landfill vehicles. At sites where excavations can be made below grade, the trench method of construction may be followed. Where this is not feasible because of topography or groundwater conditions, the area method may be practiced, resulting in a mound or hill rising above the original ground. Since no ground is excavated in the area method, soil usually must be hauled to the site from some other location. Variations of the area method may be employed where a landfill site is located on sloping ground, in a valley, or in a ravine. The completed landfill eventually blends in with the landscape.

Controlling By-products

Organic material buried in a landfill decomposes by anaerobic microbial action. Complete decomposition usually takes more than 20 years. One of the by-products of this decomposition is methane gas. Methane is poisonous and explosive when diluted in the air, and it can flow long distances through porous layers of soil. If it is allowed to collect in basements or other confined areas, dangerous conditions may arise. In modern landfills, methane movement is controlled by impermeable barriers and by gas-venting systems. In some landfills the methane gas is collected and recovered for use as a fuel.

A highly contaminated liquid called leachate is another by-product of decomposition in sanitary landfills. Most leachate is the result of runoff that infiltrates the refuse cells and comes in contact with decomposing garbage. If leachate reaches the groundwater or seeps out onto the ground surface, serious environmental pollution problems can occur, including the possible contamination of drinking-water supplies. Methods of controlling leachate include the interception of surface water in order to prevent it from entering the landfill and the use of impermeable liners or barriers between the waste and the groundwater. New landfill sites should also be provided with groundwater-monitoring wells and leachate-collection and treatment systems.

Importance in Waste Management

In communities where appropriate sites are available, sanitary landfills usually provide the most economical option for disposal of nonrecyclable refuse. However, it is becoming increasingly difficult to find sites that offer adequate capacity, accessibility, and environmental conditions. Nevertheless, landfills will always play a key role in solid-waste management. It is not possible to recycle all components of solid waste, and there will always be residues from incineration and other treatment processes that will eventually require disposal underground. In addition, landfills can actually improve poor-quality land. In some communities properly completed landfills are converted into recreational parks, playgrounds, or golf courses.

Recycling

Separating, recovering, and reusing components of solid waste that may still have economic value is called recycling. One type of recycling is the recovery and reuse of heat energy, a practice discussed separately in Incineration. Composting can also be considered a recycling process, since it reclaims the organic parts of solid waste for reuse as mulch or soil conditioner. Still other waste materials have potential for reuse. These include paper, metal, glass, plastic, and rubber.

Separation

Before any material can be recycled, it must be separated from the raw waste and sorted. Separation can be accomplished at the source of the waste or at a central processing facility. Source separation, also called curbside separation, is done by individual citizens who collect newspapers, bottles, cans, and garbage separately and place them at the curb for collection. Many communities allow "commingling" of nonpaper recyclables (glass, metal, and plastic). In either case, municipal collection of source-separated refuse is more expensive than ordinary refuse collection.

In lieu of source separation, recyclable materials can be separated from garbage at centralized mechanical processing plants. Experience has shown that the quality of recyclables recovered from such facilities is lowered by contamination with moist garbage and broken glass. The best practice, as now recognized, is to have citizens separate refuse into a limited number of categories, including newspaper; magazines and other wastepaper; commingled metals, glass, and plastics; and garbage and other nonrecyclables. The newspaper, other paper wastes, and commingled recyclables are collected separately from the other refuse and are processed at a centralized material recycling facility, or MRF. A modern MRF can process about 300 tons of recyclable wastes per day.

At a typical MRF, commingled recyclables are loaded onto a conveyor. Steel cans ("tin" cans are actually steel with only a thin coating of tin) are removed by an electromagnetic separator, and the remaining material passes over a vibrating screen in order to remove broken glass. Next, the conveyor passes through an air classifier, which separates aluminum and plastic containers from heavier glass containers. Glass is manually sorted by colour, and aluminum cans are separated from plastics by an eddy-current separator, which repels the aluminum from the conveyor belt.

Reuse

Recovered broken glass can be crushed and used in asphalt pavement. Colour-sorted glass is crushed and sold to glass manufacturers as cullet, an essential ingredient in glassmaking. Steel cans are baled and shipped to steel mills as scrap, and aluminum is baled or compacted for reuse by smelters. Aluminum is one of the smallest components of municipal solid waste, but it has the highest value as a recyclable material. Recycling of plastic is a challenge, mostly because of the many different polymeric materials used in its production. Mixed thermoplastics can be used only to make lower-quality products, such as "plastic lumber."

In the paper stream, old newspapers are sorted by hand on a conveyor belt in order to remove corrugated materials and mixed papers. They are then baled or loose-loaded into trailers for shipment to paper mills, where they are reused in the making of more newspaper. Mixed paper is separated from corrugated paper for sale to tissue mills. Although the processes of pulping, de-inking, and screening wastepaper are generally more expensive than making paper from virgin wood fibres, the market for recycled paper should improve as more processing plants are established.

Rubber is sometimes reclaimed from solid waste and shredded, reformed, and remolded in a process called revulcanization, but it is usually not as strong as the original material. Shredded rubber can be used as an additive in asphalt pavements, and discarded tires may be employed as swings and other recreational structures for use by children in "tire playgrounds." In general, the most difficult problem associated with the recycling of any solid-waste material is finding applications and suitable markets. Recycling by itself will not solve the growing problem of solid-waste management and disposal. There will always be some unusable and completely valueless solid residue requiring final disposal.

Types of Solid Waste Management

Some of the major types of solid waste management are as follows: a) Municipal Solid Waste (MSW), b) Hazardous Wastes, c) Industrial Wastes, d) Agricultural Wastes, e) Bio-medical Wastes, f) Waste Minimization.

The combined effects of population explosion and changing modern living standard have had a cumulative effect in the generation of a large amount of various types of wastes. Solid waste can be classified into different types depending on their sources:

Municipal Solid Waste

The term municipal solid waste (MSW) is generally used to describe most of the non-hazardous solid waste from a city, town or village that requires routine collection and transport to a processing or disposal site. Sources of MSW include private homes, commercial establishments and institutions, as well as industrial facilities.

However, MSW does not include wastes from industrial processes, construction and demolition debris, sewage sludge, mining waste or agricultural wastes. MSW is also called as trash or garbage. In general, domestic waste and MSW are used as synonyms.

Municipal solid waste contains a wide variety of materials. It can contain food waste like vegetable and meat material, leftover food, eggshells etc, which is classified as wet garbage as well as paper, plastic, tetra-pack, plastic cans, newspaper, glass bottles, cardboard boxes, aluminium foil, meta items, wood pieces, etc., which is classified as dry garbage. The different types of domestic wastes generated and the time taken for them to degenerate is illustrated in the table given below.

Table: Domestic wastes and their degeneration time

Common domestic wastes	Approximate time taken for degeneration
Organic kitchen waste vegetables, fruits	1-2 weeks
Paper, cardboard paper	15 days-1 month
Cotton clothes	2-5 months
Woolen clothes	about a year
Metal cans, tin, aluminium	100-500 years
Plastics	1 million years

The Functional Elements of MSW Management

The municipal solid waste industry has four components: recycling composting, land-filling, and waste-to-energy via incineration. The primal steps are generation, collection, sorting and separation, transfer and disposal/utilisation.

1. Waste generation encompasses activities in which materials are identified as no longer being of value and are either thrown out or gathered together for disposal.

2. The functional element of collection includes not only the gathering of solid waste and recyclable materials, but also the transport of these materials, after collection, to the location where the collection vehicle is emptied. This location may be a material processing facility, a transfer station or a landfill disposal site.

3. Waste handling and separation involves activities associated with waste management until the waste is placed in storage containers for collection. Handling also encompasses the movement of loaded containers to the point of collection.

Separating different types of waste components is an important step in the handling and storage of solid waste at the source. The types of means and facilities that are now used for the recovery of waste materials that have been separated at the source include curbside collection, drop off and buy back centers.

4. Transfer and transport involves two main steps. First, the waste is transferred from a smaller collection vehicle to larger transport equipment. The waste is then transported, usually over long distances, to a processing or disposal site.

5. Today the disposal of wastes by land filling or land spreading is the ultimate fate of all solid wastes, whether they are residential wastes collected and transported directly to a landfill site, residual materials from materials recovery facilities (MRFs), residue from the combustion of solid waste, compost or other substances from various solid waste processing facilities.

A modern sanitary landfill is not a dump; it is an engineered facility used for disposing of solid wastes on land without creating nuisances or hazards to public health or safety, such as the breeding of insects and the contamination of ground water. Municipal solid waste can be used to generate energy.

Several technologies have been developed that make the processing of MSW for energy generation cleaner and more economical than ever before, including landfill gas capture, combustion, pyrolysis, gasification, and plasma arc gasification. While older waste incineration plants emitted high levels of pollutants, recent regulatory changes and new technologies have significantly reduced this concern.

In USA, EPA regulations in 1995 and 2000 under the Clean Air Act have succeeded in reducing emissions of dioxins from waste-to-energy facilities by more than 99 percent below 1990 levels, while mercury emissions have been by over 90 percent. The EPA noted these improvements in 2003, citing waste-to-energy as a power source “with less environmental impact than almost any other source of electricity”.

Municipal solid waste management is more of an administrative and institutional mechanism failure problem rather than a technological one. Until now, MSW management has been considered to be almost the sole responsibility of urban governments, without the participation of citizens and other stakeholders.

The Centre and the Supreme Court, however, have urged that this issue be addressed with multiple stakeholder participation. Cities in India spend approximately 20% of the city budget on solid waste services.

Hazardous Wastes

Hazardous wastes are those that can cause harm to human and the environment.

Characteristics of Hazardous Wastes

Wastes are classified as hazardous if they exhibit any of four primary characterises based on physical or chemical properties of toxicity, reactivity ignitability and corrosively.

- Toxic wastes: Toxic wastes are those that are poisonous in small or trace amounts. Some may have acute or immediate effect on human or animals. Carcinogenic or mutagenic causing biological changes in the children of exposed people and animals. Examples: pesticides, heavy metals.
- Reactive wastes: Reactive wastes are those that have a tendency to react vigorously with air or water are unstable to shock or heat, generate toxic gases or explode during routine management. Examples: Gun powder, nitro glycerin.
- Ignitable waste: Are those that burn at relatively low temperatures (<60 °C) and are capable of spontaneous combustion during storage transport or disposal. Examples: Gasoline, paint thinners and alcohol.
- Corrosive wastes: Are those that destroy materials and living tissues by chemical reactions? Examples: acids and base.
- Infectious wastes: Included human tissue from surgery, used bandages and hypoderm needles hospital wastes.

Sources of Hazardous Wastes

- Chemical manufacturing companies, petroleum refineries, paper mills, smelters and other industries. Plastic industries thousand of chemicals are used in industries every year. When used incorrectly or inappropriately they can become health hazards.
- PCBs (Polychlorinated biphenyls) are resistant to fire and do not conduct electricity very well, which makes them excellent materials for several industrial purposes. Rainwater can wash PCBs out of disposal areas in dumps and landfills thus contaminating the water.
- PCBs do not break open very rapidly in the environment and thus retain their toxic characteristics. They cause long-term exposure problems to both human and wildlife. Many household chemicals can be quite toxic to humans as well as wildlife.

Most of the dangerous substances in our homes are found in various kinds of clearness, solvents and products used in automotive care. When these products are used incorrectly they have the potential to be harmful.

Effects of Hazardous Wastes

As most of the hazardous wastes are disposed off or in land, the most serious environmental effect is contaminated ground water. Once ground water is polluted with hazardous wastes, it is very often not possible to reverse the damage. Pesticides form residues in the soil that are washed into streams which then carry them forward.

The residues may persist in PCBs (poly chlorinated biphenyls) are concentrated in the kidneys and liver and cause damage; they cause reproductive failure in birds and mammals. The soil or in the bottom of lakes and rivers.

Exposure can occur through ingestion, inhalation and skin contact, resulting acute or chronic poisoning. Lead, mercury and arsenic are hazardous substances which can often refer to as heavy metals. Most of the lead absorbed by people is stored in the bones.

Lead can affect red blood cells by reducing their ability to carry oxygen and shortening their life span. Lead may also damage nervous tissue, resulting in brain disease. Mercury is used in production of chlorine and as a catalyst in the production of some plastics.

Mercury build up in body over long period of time is known to cause brain damage. Minamata disease occurs due to mercury poisoning. Vinyl chloride is a chemical that is widely used in plastic manufacture. A long continuous exposure in humans it can cause deafness, vision problem circulation disorders and bone deformities.

Control of Hazardous Wastes

Common methods for disposing of hazardous wastes are land disposal and incineration Industries need to be encouraged to generate less hazardous waste in the manufacturing process. Although toxic wastes cannot be entirely eliminated, technologies are available for minimizing recycling and treating the wastes.

Integrated pest management practices (IPM) reduce the usage of pesticides. Substitute the use of PCBs and vinyl chloride with chemicals that are less toxic. Polyvinyl chloride use can be lowered by reducing the use of plastics.

Industrial Wastes

These contain more of toxic and require special treatment.

Source of Industrial Wastes

Food processing industries, metallurgical chemical and pharmaceutical unit's breweries, sugar mills, paper and pulp industries, fertilizer and pesticide industries are major ones which discharge toxic wastes. During processing, scrap materials, tailings, acids etc.

Effects of Industrial Wastes

Most common observation is that the health of the people living in the neighborhood of dumping sites is severely affected. The exposure may cause disorders of nervous system, genetic defects, skin diseases and even caner.

The liquid effluents discharged by the industries contain inorganic and organic pollutants and they enter into water bodies causing destruction of fish, formation of sediments, and pollution of ground water and release of foul odours.

Control of Industrial Wastes

Waste minimization technologies have to be developed. Source reduction recycling and reuse of materials need to be practiced on a large scale. Hazardous waste should not mix up with general waste. Source reduction involves altering the design, manufacture or use of products and materials to reduce the amount and toxicity of materials that get thrown away.

Local communities and voluntary organizations should educate the industrialists as well as the public about dangers of pollution and the need to keep the environment clean. Land filling, incineration and composting technologies to be followed. Biogas is obtained from solid waste treatment of industrial and mining waste is done for the recovery of useful products.

Agricultural Wastes

Sources of Agricultural Wastes

The waste generated by agriculture includes waste from crops and live stock. In developing countries, this waste does not pose a serious problem as most of it is used e.g., dung is used for manure, straw is used as fodder. Some agro-based industries produce waste e.g., rice milling, production of tea, tobacco etc. Agricultural wastes are rice husk, degasses, ground nut shell, maize cobs, straw of cereals etc.

Effects of Agricultural Wastes

If more C:N ratio wastes like paddy husk or straw may cause immobilization of nutrients if applied on the fields. It occupies to large land areas if not properly disposed.

Management of Agricultural Wastes

- Waste to energy:
  - Gasification:
    - It is the process in which chemical decomposition of biomass takes place in the presence of controlled amounts of oxygen, producing a gas. This gas is cleaned and used in an internal combustion engine to produce electric power.
    - Without clean up also, the gas can be used in boilers to produce electric power. This technology is highly suited to generate electric power from agricultural wastes like rice husks, groundnut shells etc.
  - Pyralysis:
    - It is similar to gasification except that the chemical decomposition of biomass wastes takes place in the absence or reduced presence of O_2 at high temp. Mixtures of gases result from decomposition including H_2, NH_4Co, CO_2 depending on the organic nature of waste matter. This gas used for power generation.

- Biogas production:
  - Animal wastes, food processing wastes and other organic matter are decomposed anaerobically to produce a gas called biogas. It contains methane and CO_2. The methane can provides gas for domestic use. The byproduct of this technology is slurry, settled out the bottom of the digester. This can be used as manure.
- Agricultural waste like corn cobs, paddy husk, bagasse of sugarcane, waste of wheat, rice and other cereals, cotton stalks, coconut wastes, jute waste etc. can be used in making of paper and hard board.

Bio-medical Wastes

Bio-medical waste means any waste, which is generated during the diagnosis, treatment or immunisation of human beings or animals or in research activities pertaining thereto or in the production or testing of biological.

Segregation, Packaging, Transportation and Storage

- Bio-medical waste shall not be mixed with other wastes.
- Bio-medical waste shall be segregated into containers/bags at the point of generation prior to its storage, transportation, treatment and disposal. The containers shall be properly labelled.
- Notwithstanding anything contained in the Motor Vehicles Act, 1988, or rules there under, untreated biomedical waste shall be transported only in such vehicle as may be authorised for the purpose by the competent authority as specified by the government.
- No untreated bio-medical waste shall be kept stored beyond a period of 48 hours. Provided that if for any reason it becomes necessary to store the waste beyond such period, the authorised person must take permission of the prescribed authority and take measures to ensure that the waste does not adversely affect human health and the environment.

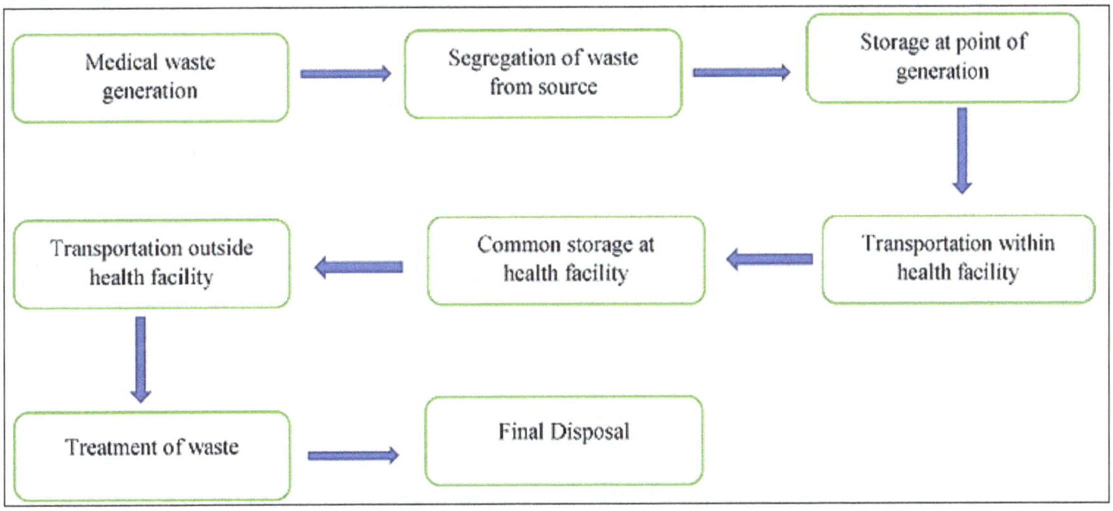

Bio-medical Waste Management Flowchart.

Categories of Bio-medical Wastes

Table: Categories of Bio-medical Wastes.

Option	Treatment and Disposal	Waste Category
Cat. No. 1	Incineration/deep burial	Human Anatomical Waste (human tissues, organs, body parts)
Cat. No. 2	Incineration/deep burial	Animal Waste Animal tissues, organs. Body parts carcasses, bleeding parts, fluid, blood and experimental animals used in research, waste generated by veterinary hospitals/colleges, discharge from hospitals, animal houses)
Cat. No. 3	Local autoclaving/micro waving/ incineration	Microbiology and Biotechnology waste (wastes from laboratory cultures, stocks or specimens of micro-organisms live or attenuated vaccines, human and animal cell culture used in research and infectious agents from research and industrial laboratories, wastes from production of biological, toxins, dishes and devices used for transfer of cultures)
Cat. No. 4	Disinfections (chemical treatment/autoclaving/micro waving and mutilation shredding	Waste Sharps (needles, syringes, scalpels blades, glass etc. that may cause puncture and cuts. This includes both used and unused sharps)
Cat. No. 5	Incineration/destruction and drugs disposal in secured landfills	Discarded Medicines and Cytotoxic drugs (wastes comprising of outdated, contaminated and discarded medicines)
Cat. No. 6	Incineration, autoclaving/micro waving	Solid Waste (Items contaminated with blood and body fluids including cotton, dressings, soiled plaster casts, line beddings, other material contaminated with blood)
Cat. No. 7	Disinfections by chemical treatment autoclaving/micro waving and mutilation shredding.	Solid Waste (waste generated from disposable items other than the waste sharps such as tubing, catheters, intravenous sets etc.)
Cat. No. 8	Disinfections by chemical treatment and discharge into drain	Liquid Waste (waste generated from laboratory and washing, cleaning, housekeeping and disinfecting activities)
Cat. No. 9	Disposal in municipal landfill	Incineration Ash (ash from incineration of any bio-medical waste)
Cat. No. 10	Chemical treatment and discharge into drain for liquid and secured landfill for solids	Chemical Waste (chemicals used in production of biological, chemicals, used in disinfect ion, as insecticides, etc)

Waste Minimization

Waste production can be minimized by adopting the 3 R's principle: Reduce, Reuse, and Recycle.

- Reduce the amount and toxicity of garbage and trash that you discard.
- Reuse containers and try to repair things that are broken.
- Recycle products wherever possible, which includes buying recycled products i.e., recycled paper books, paper bags etc.

These are processes that involve integrated waste management practices (IWM). They can reduce the wastes generated by approximately 50%.

Reduce

Waste prevention, or "source reduction," means consuming and discarding less, is a successful method of reducing waste generation. Backyard composting, double sided copying of papers, purchasing durable, long- lasting environmentally friendly goods; products and packaging that are free of toxics, redesigning products to use less raw material production and transport packaging reduction by industries are the normal practices used and have yielded substantial environmental benefits.

Source reduction prevents emissions of many greenhouse gases, reduces pollutants the need saves energy, conserves resources, and reduces wastes for new landfills and combustors. It reduces the generation of waste and is generally preferred method of waste management that goes a long way toward saving the environment.

Reuse

Reuse is the process, which involves reusing items by repairing them, donating them to charity and community groups, or selling them. Reusing products is an alternative to recycling because the item does not need to be reprocessed for its use again. Using durable glassware, steel using cloth napkins or towels, reusing bottles, reusing boxes, purchasing refillable pens and pencils are suggested.

Recycling

The process of recycling, including composting, has diverted several million tons of material away from disposal. Recycled materials include batteries, recycled at a rate of 93%, paper and paperboard at 48%, and yard trimmings at 56%. These materials and others may be recycled through drop off centers, buy-back programs, and deposit systems.

Recycling prevents the emission of many greenhouse gases that affect global climate, water pollutants, saves energy, supplies valuable raw materials to industry, creates jobs, stimulates the development of greener technologies, conserves resources for our children's future, and reduces the need for new landfills and combustors. For example, by recycling of solid waste in 1996, the United States prevented, the release of 33 million tons, of carbon into the air roughly the amount emitted annually by 25 million cars.

Recycling can create valuable resources and it generates a host of environmental, financial, and social benefits. Materials like glass, metal, plastics, and paper are collected, separated and sent to processing centers where they are processed into new products.

The advantages of recycling are it conserves resources for future generation, prevents emissions of greenhouse gases and pollutants, saves energy, supplies valuable raw, materials to industries, stimulates the development of greener technologies, reduces the need for new landfills and incinerators.

LIQUID WASTE MANAGEMENT

Dealing with liquid wastes is one of the biggest challenges of urban sanitation. Liquid waste management requires capital investment, skilled personnel, coordination between different government departments and organisations, and awareness of the issues by the public. Failure to manage liquid wastes, especially human excreta, leads to health and environmental problems.

In urban Ethiopia the vast majority of people (80%) use pit latrines for human waste disposal (Central Statistical Agency, 2014). In some parts of Addis Ababa, liquid waste from waterborne toilet systems is collected via a sewerage system and sent to the wastewater treatment facility at Kality. There are, however, many houses and institutions with waterborne toilet systems that are not connected to the sewerage system. In these places, the sewage is sent to septic tanks. These different sanitation technologies require different techniques for managing and disposing of the wastes.

Pit Emptying

The process of pit emptying is sometimes called desludging. There are manual and mechanical methods for desludging, but the manual removal of faecal sludge from pit latrines poses severe risks to those undertaking the task.

Whether using manual or mechanical methods, the personal safety of anyone employed in pit emptying should be of primary importance. Operatives should wear gloves, masks and protective clothing.

Vacuum Trucks

Vacuum trucks are vehicles equipped with a storage tank and pump with a hose that is lowered into the pit to suck the sludge up and out into the storage tank. The sludge can then be easily transported to a suitable disposal site. Vacuum trucks are quick, powerful and efficient, but they are large vehicles and so access to the pits can be a problem. The size of the truck can limit their use in areas where roads are narrow and twisting. Truck operators will charge a fee for their services, which is another factor that needs to be considered.

Vacuum trucks are used to empty pit latrines and septic tanks.

Vacutug

The Vacutug, shown in figure, is basically a smaller version of the vacuum truck. It was devised by UN-Habitat as a system that could replace manual emptying. The Vacutug is a mechanical system that can be manufactured locally using readily available components. It is affordable, easily serviceable, able to operate in narrow passageways where vacuum trucks cannot go and is capable of sucking out waste sludge for transportation to a larger tanker vehicle. It can empty pits down to 2 m deep.

The Vacutug.

Hand-operated Pumps

Even the Vacutug cannot get everywhere. The Sludge Gulper, like the one shown in figure, is an example of a smaller hand-operated pump. These can be taken to pit latrines that are inaccessible to larger pumps. It is a simple design consisting of a PVC pipe containing two valves and can be built using locally available materials. The sludge is pumped up by hand, collected in a container and taken away for disposal. Care is needed to ensure that the operator and other helpers do not come into contact with the sludge and that it is not spilled.

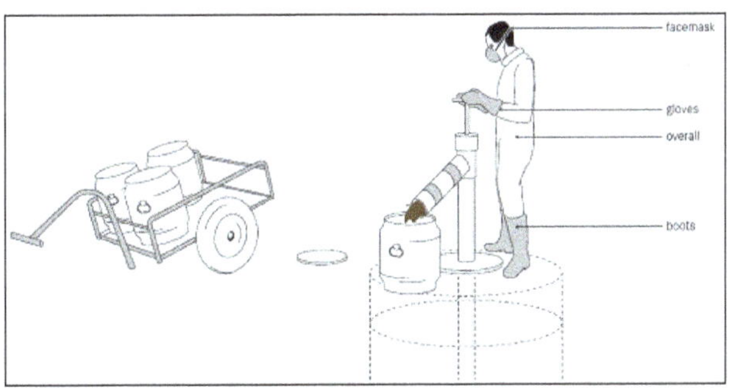

Hand-operated sludge pump. The pump is located over the pit and the operator (wearing suitable protective clothing) pumps the handle up and down. Sludge rises up the vertical pipe and is expelled into the container.

Disposal of the Sludge

Several options are available for disposal of the collected sludge. It can be put directly onto land and used as a soil conditioner, but this is only possible if it has been left untouched for at least two years. Fresh, untreated wet sludge poses high risks for human health and so should not be put on land used to grow crops.

Drying the sludge will kill most pathogens. This can be achieved using drying beds, where sludge is put into shallow tanks to a depth of about 300 mm. The base of the tank is sloped and covered with a layer of sand (forming a 'bed') to allow liquid to drain out of the sludge. In the warm climate of Ethiopia and without rain, after about a week the sludge will be dry enough to be lifted by a shovel.

Sludge in drying beds.

The sludge can also be composted by mixing it with vegetable matter, or biogas can be obtained by anaerobic digestion. Whichever method is used, faecal sludge disposal must be carefully managed and operated in order to ensure that the associated risks to health and the environment are avoided.

Septic Tanks

A septic tank is an underground and watertight tank made of concrete, fibreglass or PVC in which sewage is collected and partially treated. These are used for water-flushed systems that are not connected to a sewer. Concrete septic tanks are commonly used in Ethiopia.

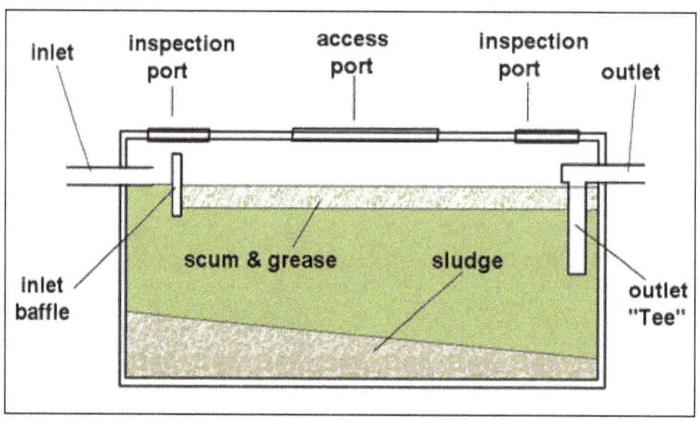

Diagram of a septic tank.

Wastewater enters the tank, remains there for a time, and is displaced out of the tank by new wastewater coming in. There are no pumps or mechanical parts. The time the wastewater remains in the tank is called the retention time and should be a minimum of 24 hours. In this period, the solid matter in the sewage settles to the bottom of the tank, where it is partially degraded by anaerobic micro-organisms.

The liquid above the sludge is relatively free of solids, but it does contain dissolved organic and inorganic chemicals that are not treated. Light substances such as oil and grease form a scum and float to the top. The position of the outlet ensures that only water from the middle of the tank is displaced outwards. This effluent is disposed of to a soakaway (also called a soak pit or seepage pit) or to a drainfield.

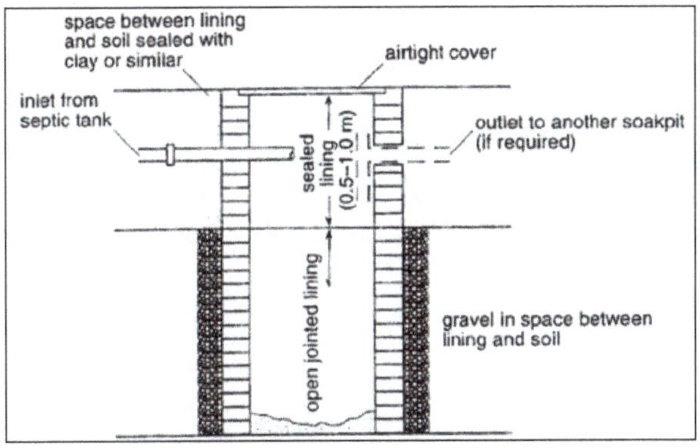

Diagram of a soakaway.

The soakaway is a covered, unsealed pit lined with bricks or stone. The wastewater from the septic tank seeps into the soil through the base of the pit and through the spaces in the lining material.

A drainfield is a field where the effluent from a septic tank is disposed of. It can be constructed by laying a network of perforated pipes (pipes with tiny holes) in excavated soil which has a bed of gravel. Soil is placed over the top of the perforated pipe, and plants are grown on top of the drain-field. The wastewater is sent into the perforated pipe and percolates (seeps) through the gravel and goes into the ground, replenishing the groundwater.

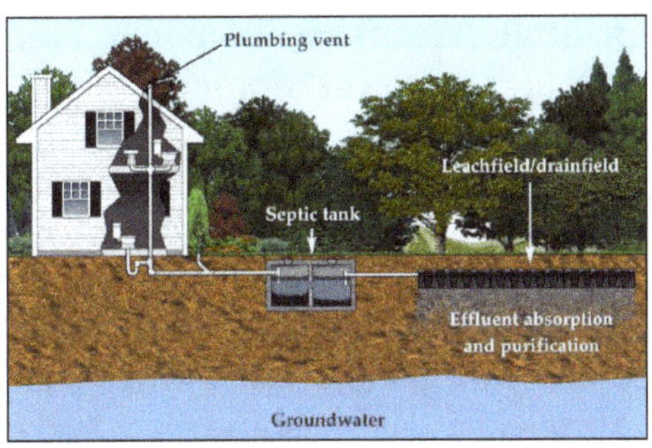

Diagram showing the location of a septic tank and a drainfield.

Septic tanks need to be desludged when the sludge depth is approximately two-thirds the depth of the tank. You can estimate this by putting a graduated stick into the tank. Depending on usage, this is typically once every 12 months. A vacuum truck is usually used to suck the sludge out for disposal.

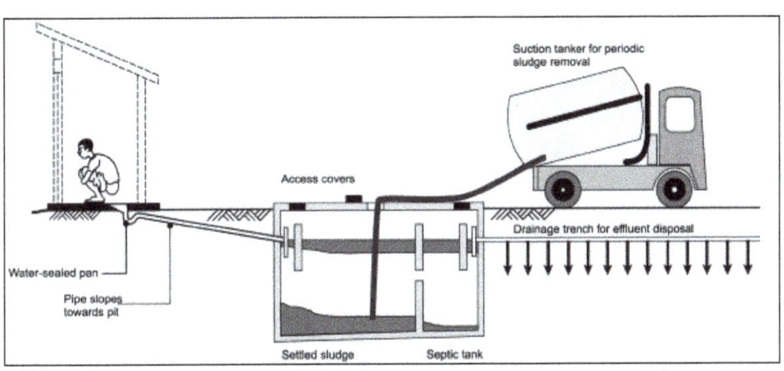

Diagram of a septic tank being desludged by vacuum truck.

If a sewer network is later installed in an area where there are septic tanks (as often happens in developing countries), the sewage flow can be connected directly to the sewer, thus bypassing the septic tank. This allows for better treatment (and possibly reuse) of the sewage, and of course eliminates the need for a septic tank.

Wastewater Treatment Options

Septic tanks remove solids from sewage, but treatment is minimal and the effluent contains high levels of dissolved organic matter and ammonia. Other options for better treatment are available.

The aim in wastewater treatment is to:

- Reduce the amount of biodegradable material and solids.
- Remove toxic materials.
- Eliminate pathogenic micro-organisms.

There are several ways of treating sewage and other wastewaters with different costs and requirements. One that is particularly effective and economical in warm regions of the world is waste stabilisation ponds.

Waste Stabilisation Ponds

Waste stabilisation ponds are natural or constructed ponds used for treating sewage or other wastewaters biologically by harnessing the power of sunlight and wind. They are ideal for tropical countries like Ethiopia. The sewage collected by the sewerage network in Addis Ababa goes to waste stabilisation ponds at Kality.

In a typical waste stabilisation pond system, effluent that has passed through a screen is sent through a series of ponds with a total retention time of between 10 and 50 days. Bacteria in the ponds degrade the organic waste and work symbiotically with algae, which provide oxygen through photosynthesis. (Symbiotically means in a relationship where two types of living organism live together for their mutual benefit). Oxygenation also occurs through the action of wind and by diffusion from the air. No mechanical equipment is used in the ponds, so operation and maintenance costs are very low. Land requirement is, however, high.

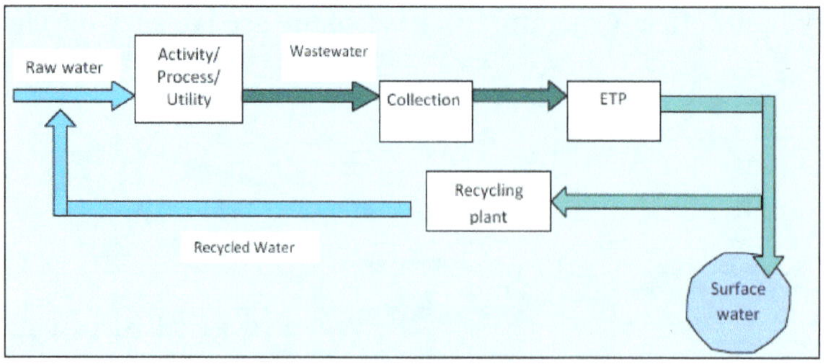

Waste water recycling.

The major part of the biodegradation of the sewage takes place in the facultative ponds. Facultative ponds are 1–1.5 m deep, with a retention time of between 5 and 30 days. In these ponds the upper layers are aerobic and the lower layers of water are anaerobic. Solids settle to the bottom and are anaerobically digested, so sludge removal is rarely needed.

Maturation ponds are placed after facultative ponds for the purpose of pathogen reduction. These are usually 0.5–1.5 m deep with a retention time of between 15 and 20 days. These ponds serve to inactivate pathogenic bacteria and viruses through the action of UV radiation from sunlight and the greater algal activity in these shallow ponds, which raises the pH above 8.5. (pH is a measure of acidity and alkalinity. It has a scale from 0–14: pH 7 is neutral, less than 7 is acid and more than 7 is alkaline). The long retention time in the maturation ponds also enhances the sedimentation of the eggs of intestinal parasitic worms.

A waste stabilisation pond – the green colour of the pond due to algae.

If the wastewater has a very high level of pollutants, anaerobic ponds can be used ahead of the facultative ponds. Anaerobic ponds are 2–5 m deep and are nearly devoid of oxygen. Their retention time is one to seven days. Solids settle to the bottom, forming a sludge, and anaerobic digestion takes place, producing methane. Up to 60% of the organic material can be removed in this process.

To prevent sewage from leaching away, and to preserve the effluent for reuse later, the ponds should have a liner. This can be made of clay, asphalt, compacted earth or any other impermeable material (material that does not let fluid pass through). To prevent run-off from entering the ponds and to prevent erosion, a protective raised earth barrier can be constructed around the ponds using the excavated material from their construction. Finally, a fence is needed to keep people and animals out.

Any scum that builds up on the surface of the facultative and maturation ponds should be removed to allow sunlight to reach all the algae and also to increase surface aeration. Large plants that are present in the water should be removed.

Treated sewage can be reused in crop irrigation if safe limits of faecal bacteria and intestinal parasite eggs are achieved in the treatment process. At the same time as treating wastewater, pond systems have been used to increase protein production through the rearing of fish (such as Tilapia) and ducks in maturation ponds.

Reed Beds

Reed beds, or artificially constructed wetlands with emergent plants, have been used to treat sewage in many parts of the world. In Ethiopia, studies have been carried out on this type of wastewater treatment system for domestic wastewaters in Addis Ababa and Kombolcha, and brewery effluent in Addis Ababa.

Reed beds are ideal for warm countries where plants grow rapidly. They have low operational costs but they do require a lot of land.

The plants (usually reed species Phragmites australis or Phragmites communis) are grown in rows in beds of soil or gravel lined with an impermeable clay or synthetic liner. The effluent requiring treatment is fed into the bed, which typically has a depth of 600 mm. The base of the reed bed has a slope to enable collection of the effluent after treatment.

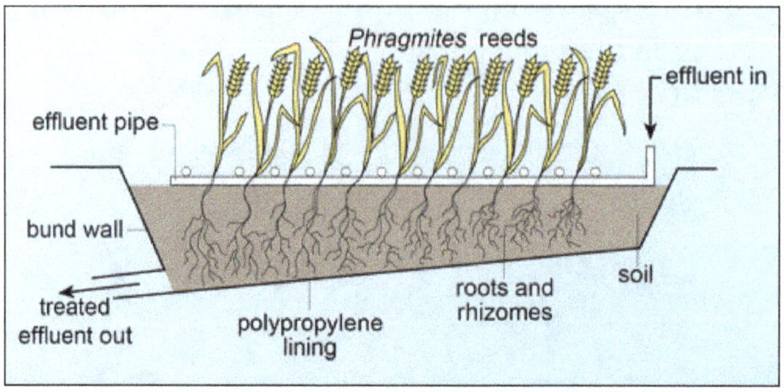

The reed bed system. When full-grown, the reeds are between 1.5 and 3.0 m high.

The effluent is distributed through pipes and nozzles onto the reed bed and then percolates down to the roots and rhizomes (horizontal underground stems) of the reeds. The root and rhizome system provides a mix of aerobic and anaerobic conditions that encourage a diversity of microbial species in the soil. As a result, the reed bed system has potential for treating a wide range of pollutants. For example, although micro-organisms that are capable of biodegrading many synthetic chemicals (such as some common pesticides) are found in soil, they not normally present in effluent treatment plants, so reed beds can be effective for treating effluents that contain these types of chemicals.

Mechanical-biological Wastewater Treatment

Wastewater treatment can be undertaken using a sequence of processes in a mechanical-biological system. This treatment is faster than using natural systems such as waste stabilisation ponds or reed beds and requires less space. These factors make it desirable for sewage treatment in towns with large populations where there is not enough land for natural systems. However, mechanical-biological systems are more expensive because of the equipment required and the need for skilled personnel to operate them.

A mechanical-biological wastewater treatment plant.

These systems typically have three main stages: preliminary, primary and secondary treatment.

Preliminary Treatment

In this first stage screens remove large items such as pieces of wood, metal, rags, paper or plastic that have got into the sewerage system. Removing them protects the structures and equipment in the wastewater treatment plant. Paper and rags in the wastewater flow are sometimes shredded by comminutors, which are rotating, slotted drums equipped with cutting blades. The shredded material can then be returned to the flow further downstream without causing harm.

Screening of wastewater. Solid materials are retained on the screen while the liquid passes through.

Small stones and grit have to be removed from the flow. Grit is comprised of very small pieces of sand, stone and possibly also glass and metal. All these materials can increase the rate of wear of mechanical equipment and can also settle easily in pipes, causing blockages. The grit settles out in grit channels and can be removed daily by manual or mechanical means and dumped at a landfill site.

A grit channel that slows the rate of flow and allows grit to settle out by gravity.

Primary Treatment

In the primary treatment stage, fine solids in the wastewater are removed by settlement in a sedimentation tank. A properly designed and well-operated primary sedimentation tank will reduce the suspended solids content of the wastewater by between 50 and 70%, and the biochemical oxygen demand (BOD) by between 25 and 40%.

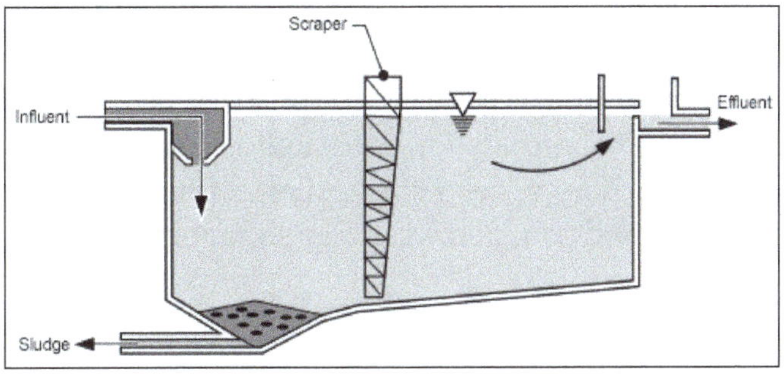

Diagram of a rectangular sedimentation tank.

Secondary Treatment

This is the biological stage of treatment. In secondary treatment the organic matter in the sewage is biodegraded by micro-organisms using oxygen. Oxygen levels are increased artificially by various means to ensure removal of organic matter. Also in this stage, ammonia in the sewage is converted to nitrate. This is followed by a second sedimentation stage to remove solids produced by the microbial activity. The treated effluent should be clear, free of pathogens and safe to discharge into a river or possibly be reused for irrigation.

Sullage Management

Many people simply throw sullage onto the ground or into the street. In small quantities this may be acceptable, but in densely populated urban areas proper handling and disposal is required. Some of the disadvantages of improper disposal of sullage include the potential to contaminate the soil, pollute water sources and create favourable breeding conditions for disease vectors.

Sullage can be discharged to sewers or septic tanks in areas where these facilities exist. Where they do not, it is necessary to construct a pit near the household to dispose of sullage properly. The pit should be filled with gravel or sand. A sullage pit keeps the wastewater in one place and encourages it to soak quickly into the ground.

Pit for disposal of sullage.

AGRICULTURAL WASTE MANAGEMENT

The first goal of any waste management system is to maximize the economic benefit from the waste resource and maintain acceptable environmental standards. To be practical, the system must also be affordable and suitable to the operation. If wastes are not properly handled they can pollute surface and groundwater and contribute to air pollution. Most people think of manure first when they think of farm waste. While manure is an important component, farm waste in a livestock operation can also include waste forage, dead stock, silage effluent and milkhouse waste. In horticultural operations, culls, diseased product, wash line sediment and processing plant wastes are common by-products.

In addition to these, all farm operations generate plastic waste material ranging from silage wrap to pesticide or drug containers.

Management that puts into practice the principles of the four Rs of Reduce, Reuse, Recycle and Recover is the best first option:

- Reduce the amount of waste product generated;
- Reuse the waste product on the farm or provide it for others to use;
- After reducing and reusing as much of the waste product as possible, recycle the product either on-farm, such as with land application of manure, or off-farm, such as with plastic recycling programs.
- Recover methane gas from manure waste.

Only after considering the four Rs should farm waste be disposed of.

Farm Waste By-product as a Resource

Many farm by-products can be economically valuable resources when managed correctly. Manure, for example, is a valuable resource because of its fertilizing and soil conditioning properties. Horticultural washwater can be economically recycled. Farm plastics can be recycled or reused. If systems for storage and handling are substandard these wastes can degrade the environment on and off the farm.

Relevant Guidelines and Regulations

Farmers should be aware of the environmental guidelines and regulations which apply to farm operations. Best Management Practices integrate principles of production, business goals, sustainability and environmental quality in farm resource management systems.

Challenge

Our society is increasingly concerned with the environmental consequences of all activities. Farming operations are no exception. Our dependence on groundwater, the delicate balance of our coastal estuaries and the economic importance of tourism each provide ample reason to use the best management practices to handle wastes.

Green Consumerism

Accepting the environmental challenge and projecting a public image of good environmental stewardship can provide producers with a competitive advantage. While governments and international bodies propose formal solutions, consumers are encouraging changes in the marketplace. "Green consumerism" is a growing trend that is becoming an increasingly important factor at home and in many of the countries that are markets for Canadian agricultural products. Powerful consumer actions have the potential to affect the price and marketability of products. In a broader economic context, it is not acceptable to be competitive in the global marketplace at the expense of the environment.

Water

All water for human consumption and most water for other purposes comes from groundwater.

Because the upper layer of soil is generally thin and the underlying bedrock aquifer is extensively fractured, all areas on the Island are susceptible to groundwater contamination. Fish and wildlife are dependant on clean surface water resources and their abundance contributes to the Island economy through tourism, sportfishing, hunting, trapping and wildlife observation. Coastal estuaries have some of the most productive shellfish grounds in North America.

Climate

Agricultural activities both absorb and produce "greenhouse" gases. Gases such as carbon dioxide, methane and nitrous oxide block the escape of heat energy and produce a warming trend in the earth's atmosphere. Crop growth requires carbon dioxide while animal production and vehicle operation emit carbon dioxide. Improved treatment, handling and utilization of manure offers the greatest potential for the reduction of these gases from agricultural sources.

Livestock and Poultry Waste Management

Manure Management

The increasing size of farm operations and the expanding residential land use in rural areas has greatly increased environmental concerns over nuisance odours and the potential for water pollution. With good manure management practices, proper storage facilities, and adequate separation distances between non-compatible land uses, most environmental problems can be avoided. Manure management encompasses manure collection, storage, transport and land application. The goal of manure management must be to maximize the soil amending value of manure and minimize the potential for environmental degradation.

Nature of the Resource

Manure and contaminated runoff water are valuable sources of fertilizer and organic matter for soil. Manure is a dynamic organic material, continually undergoing biological and chemical changes. The value of manure as a fertilizer depends on the quantity and form of nutrients present when it is applied to land. Each phase of management may result in losses of, or changes to, the beneficial nutrients in the manure. Manure includes the faecal and urinary wastes of livestock and poultry, plus materials such as bedding and added water. The combined moisture level of faeces and urine ranges from 75% in poultry manure to 85% for swine manure. Depending on the amount of water or bedding added, manure can be solid, semi-solid or liquid.

Manure contains about 75% of the nutrients fed to livestock including nitrogen, phosphorus and potassium. Animals use only about 25% of nutrients and excrete the rest. About 50% of nitrogen and 75% of potassium in manure is found in the liquid portion. Therefore, it is important to contain the liquids for land application. Almost all the phosphorus is in the solids. When manure is diluted by water, nutrient concentrations are reduced.

Environmental Issues

Manure management practices have the potential to degrade the surrounding air and water. Odours are an unavoidable consequence of animal production and are the most apparent problem associated with manure. Minimizing problems associated with odours requires respect for individuals, in addition to good management practices.

Proper manure storage will minimize the impact on the natural environment.

Manure has potential to pollute surface water and groundwater through:

- Direct animal access to waterways.
- Runoff from manure stockpiles, barn yards and feedlots.
- Seepage from manure storage areas.
- Overflow from storage areas.
- Runoff from fields where manure has been applied.
- Runoff from pastures.

Surface water problems which may result include:

- Harmful effects on fish from oxygen depletion in streams, ponds, and estuaries.
- Nutrient enrichment of water systems due to increased levels of nitrates and phosphates which can cause eutrophication (algae blooms) of surface waters.
- Human and animal health hazards including high concentrations of bacteria in shellfish.

Manure and associated wastewater can be farm liabilities if they are not handled properly. Potential liabilities include:

- Bacterial and nitrogen contamination of water supplies.
- Potential disease in humans and livestock due to pathogenic bacteria.
- Dangerous gases produced in manure storage in the absence of oxygen. These gases include ammonia; methane, which is odourless; and hydrogen sulphide, which smells like rotten eggs. High concentrations of these gases can be harmful to the health of animals and humans.
- Nuisance to neighbours due to odours from manure.

Manure Handling and Storage

Livestock manure is classified as either a solid, semi-solid or liquid using the following criteria:

- Solid - Contains greater than 20% solids. Bedding material contributes to the solids content of the manure. It can be stacked and handled by any equipment that will move bulk materials.

- Semi-solid - (also referred to as slurry) - Contains 5% to 20% solids. Semi-solid manure is produced in livestock housing systems where limited bedding is supplied. The resulting semisolid does not flow as readily as liquid manure, nor can it be piled like solid manure.

Liquid manure storage - circular concrete tank.

- Liquid - Contains less than 5% solids. The additional liquid comes from washing and spillage from watering systems. When agitated, liquid manure can be pumped or moved by gravity flow. Milkhouse washwater and other types of wastewater are often added to the liquid manure. Manure which includes bedding or waste feed will require dilution if it is to be handled as a liquid.

The moisture content of the manure determines the type of handling and storage system. Most new swine and dairy operations use liquid systems, while the majority of beef and poultry producers have solid manure handling systems.

Storage

A manure storage facility which is of sufficient size reduces the chance of pollution from spills and allows land application to take place when soil is dry, when crops require nutrients, and when work schedules permit.

- Manure storage should be large enough to store manure, bedding, wasted feed, precipitation and all liquids for at least 210 days. A one year storage capacity is optimal.
- On PEI, the required volume of open manure storages and confinement yards will have to be increased by 0.6 cubic metres/sq metre (2 cubic feet/sq ft) of surface area to allow for precipitation.
- Proper management of all liquids is essential for effective and economical manure handling and storage. Since all water which comes into contact with manure must be handled as a waste, the key to efficient management is to minimize that contact.
- Surface runoff should be diverted away from livestock and manure storage areas.
- Runoff from solid manure storage and exercise yards, milking centre washwater, silo seepage and livestock housing washwater must be stored and properly handled to ensure that groundwater, streams and other surface waters are not polluted.

Solid manure storage - curbed concrete slab with ramp.

In barn storage - manure pack.

Solid Manure

There are three common and acceptable ways of storing solid manure. These are related to the kind of livestock or poultry housing system in use. Farmers should consider animal density and roof costs versus the cost of runoff collection systems when planning a solid manure storage system.

- In Barn (solid manure pack) - Manure can be stored where produced, in confined, beddedpack housing systems. These are most commonly used for dairy and beef cattle. Dry manure poultry housing systems also store the manure where poultry are housed.
- Curbed concrete slab with runoff retention - Manure is removed and stored on a curbed concrete pad with a runoff containment system. Manure is usually moved by a tractor with a scraper blade, a front-end loader, a stable cleaner and elevator/stacker or by a ram/piston pump/air mover system.
- Curbed concrete slab with roof - Manure is removed to a roofed storage area with a concrete floor and partial sidewalls constructed of reinforced concrete.

Field Storage of Manure

Field storage of manure is a practice sometimes used in conjunction with reduced storage capacity at the barn. This practice is generally not recommended due to the high permeability of PEI soils and the fractured nature of the underlying bedrock.

Solid or semi-solid manure storage - concrete slab with sidewalls and drive-in ramp.

As well, frozen ground during the winter months can increase the risk of runoff. Loss of nutrient content of the manure can also be an important consideration. Where the practice of field storage is used, a number of precautions should be employed. Field piles should not be located within 300 metres (984 ft) of a public water supply or within 90 metres (300 ft) of a watercourse, natural wetland or residential well. Manure piles should not be located in areas subject to accumulated surface runoff or where flooding can occur. Discharge of contaminated runoff to road ditches should not be permitted.

Semi-solid Manure

There are two common and acceptable ways of storing semisolid manure:

- Curbed concrete slab with earthen banks - This type of storage requires a sloped concrete floor, concrete curbs and ramp to allow easy tractor access. Earthen sidebanks must be properly designed and constructed to prevent seepage. Environmental approval will require certification by a qualified engineer. The sloped floor allows the liquid portion of the manure to flow to the lowest point, where it can be removed by pumping. The remaining solids can be removed by a tractor fitted with a front-end loader.
- Concrete storage - Where soils are low in clay content, semisolid manure may best be stored in a roofed structure with reinforced concrete sidewalls on three sides. A concrete floor sloping downward from the open side is required to contain drainage of the liquid portion. The floor should be sealed at the walls to prevent seepage.

Liquid Manure

Liquid manure storage - underbarn concrete storage with slatted floor (under construction).

All liquid manure storages must have some type of impermeable enclosure, including concrete tanks, above-ground glass-lined steel tanks and earthen ponds. These storage systems can be covered or open. Liquid manure storages are most common in confined swine operations and free stall dairy systems. Common types of liquid manure storage on PEI are:

- Underbarn Concrete Storage - Rectangular tanks with reinforced concrete walls sealed to a concrete floor; may be located below a slatted barn floor. Toxic and explosive gases may be produced when manure is agitated prior to removal, so barns should be well-ventilated.
- Exterior Concrete Tank - Circular or rectangular tanks with reinforced concrete walls and floors; may be partially or entirely in-ground. Covers may be installed to reduce odours, to keep out precipitation or to ensure safe operation. Covers chosen to reduce the strong odours common to liquid storage can include temporary floating straw crusts, tarpaulins, plastic domes or permanent steel, wood or concrete structures. For safety reasons, in-ground or partially in-ground storage outside the barn must be fenced or have a reinforced concrete cover which will support vehicle traffic. The floor elevation of the storage must be 0.5 metres (1.6 ft) above the maximum water table and bedrock elevation.

- Earthen Lagoons - Generally, PEI soils are too permeable to consider this option without the installation of a liner. Earthen lagoons are not as environmentally reliable due to the risk of puncturing the liner during cleanout. The liner must have a permeability rating of 10-7 cm/sec. Environmental approval will require certification by a qualified engineer. Soils must be tested to determine their permeability. Other suitable liners include bentonite and geotextile materials. The floor elevation of the storage must be 1 metre (3.3 ft) above the maximum water table and bedrock elevation.

Safety

Safety design features and signage are especially important for liquid storages. They should include these measures:

- Safety Fences/Walls - a permanent safety fence or wall at least 1.5 metres (5 ft) in height should protect open liquid storages without fixed covers. This discourages access, particularly by children or livestock. Fences should be chain-link type.
- Concrete Liquid Manure Storage Covers - should be designed to support tractor loads if the tank cover is close to ground level. To avoid accidental access, the tank should be at least 0.6 metres (2 ft) above ground if the top is not designed for access.
- Locking Devices for Covers - should be used.
- Signage - all access points must be marked with suitable safety signs.

Farm operators should also observe the following safety practices:

- Never enter a liquid manure tank without a self-contained breathing apparatus coupled with a 3-person buddy system and a lifeline. The gases generated by liquid manure can be toxic and suffocating. Ventilation when agitating and pumping a manure tank is essential.
- Open flame should not be allowed near a liquid manure tank. The methane gas produced by liquid manure is highly explosive.

Planning Changes to Manure Handling and Storage System

Curbed concrete manure storage with safety fence.

Making a change in the way manure is handled is usually expensive because it often requires a fundamental change in the way many other things are done on the farm. When planning changes to your manure handling and storage system or constructing a new system, consider the following basic elements:

- Storage capacity for at least 210 days.
- Safety concerns.
- Comparative cost of manure handling systems.
- Labour efficiencies.
- The quality and adaptability of the current equipment.
- Requirements for new equipment.
- Flexibility in the system and location for future expansion.
- Moisture content of manure.
- The location of the storage in relation to neighbours, streams, wells and groundwater.
- The type of storage in relation to groundwater table and soil conditions.
- Method of collection and barnto-storage transfer.
- Potential nutrient losses.
- Application method.
- Requirements to handle other liquids such as milkhouse washwater and bathroom wastewater.
- Preventing water pollution.
- Minimizing odours.

Runoff Containment

Runoff management consists of two elements: decreasing the amount of water being contaminated by livestock operations and ensuring that contaminated runoff is contained so that it will not degrade the surrounding environment.

Farm with urban encroachment.

- Every effort to reduce the volume of runoff coming onto a livestock site will pay off in reduced storage size and costs. All runoff should be diverted away from livestock housing and manure storage areas. Perimeter diversion ditches, berms and dykes, and grassed or paved waterways can all be effective depending on the topography of the site.
- Roofed exercise yards will eliminate runoff.
- The use of eaves troughs on all roofs will allow roof drainage to be controlled and diverted.
- Contaminated runoff from manure storage and livestock areas must be contained and handled as part of the manure handling system. Runoff may be added to an existing liquid

manure storage provided the storage has the capacity to handle the additional volume. Excess water in liquid manure storage does increase hauling and spreading costs. It also tends to hinder the formation of a surface crust, resulting in increased nitrogen losses and odour generation.

- Where solid manure systems are used, runoff must be handled separately. Runoff should be diverted to a separate liquid storage system. The required storage volumes will depend on local precipitation and the size and surfacing of the livestock area.
- Contaminated runoff can be treated in a constructed wetland.

Setback Considerations

Adequate separation between livestock facilities and neighbours is one means of compensating for normal odour production, reducing the potential for nuisance conflicts. Proximity to developments can determine the potential for future growth of the operation. Greater separation distances afford more opportunity for odours to become diluted by mixing with the air. When evaluating sites for new operations only, you must select a location that will impact on as few neighbours as possible. Proposals for all new or expanded livestock operations are reviewed by the Department of Fisheries, Aquaculture and Environment.

The recommended minimum separation distance (MSD) between a livestock operation and a single residence or residential and recreational areas varies with the following factors:

- Size of the agricultural operation measured in animal units.
- Degree of expansion from existing operation.
- Type of manure storage.
- Type of housing.
- Type of livestock.

In general, larger separation distances are recommended as the size of the operation increases. Municipalities may require different siting criteria from those recommended here. The location of new operations must always be cleared with municipal authorities. Municipal property maps are very useful for evaluating new sites.

Watercourses, Wetlands and Wells

Plan the location of livestock facilities and manure storage to maximize the separation distance from watercourses, wetlands, and wells. This is particularly important with earthen storages and in areas where the groundwater table is shallow or where bedrock is found close to the surface of the ground.

Required Minimum Separation Distance Between Manure Storage and Watercourses, Wetlands, and Wells		
Storage Type	Distance	
	Watercourse or Wetland, m (ft)	Source of Domestic Water, m (ft)
On-farm Storage Facility	90 (300)	90 (300)*

Field Storage	90 (300)	90 (300)*
Composting	90 (300)	90 (300)*
		* public water supply 300 m (984 ft)

- Wells should be located uphill from storages and constructed in a manner that will reduce the risk of pollutants entering the well.
- Grouting the annular space outside the casing with cement or bentonite grout must be carried out.

Odour Management in Barns and Manure Storage

Areas Odour is a part of livestock farming. Odours from livestock facilities and manure storage and handling have the most public impact. The best time to incorporate odour management considerations is prior to the construction of new livestock buildings. Separation distance is the single most important element in avoiding odour conflicts with neighbours. When manure management systems are properly designed and operated, nuisance odours can be reduced.

Distance from neighbours and non-agricultural land use will determine the level of technology and management required to minimize nuisance odours. The most common and effective odour control methods are based on reducing the amount of odour-causing gases produced and released to the atmosphere and dispersing odours as quickly as possible.

What Causes Odours

The biological breakdown of manure produces ammonia, hydrogen sulphide and other compounds such as mercaptans and amines. Combinations of these compounds can produce offensive odours at very small concentrations (parts per billion). The types of compounds produced depend on the biological processes which take place in the manure.

Concrete liquid manure storage, with cover for odour control.

The following factors control these processes:

- Bacteria which are found in manure are responsible for creating odourous gases as they break down organic material. Aerobic bacteria, which require oxygen to survive, produce mostly carbon dioxide which is essentially odourless. Anaerobic bacteria, which thrive in the absence of oxygen, tend to produce odourous compounds such as ammonia and hydrogen sulphide. The type of bacteria present may vary at different locations throughout the manure handling system. Generally, aerobic bacteria are located near the surface, while anaerobic bacteria are beneath the surface.

- Temperature controls the rate of bacterial action. The higher the temperature, the faster the biological action and therefore the greater the gas production. This explains the fact that fewer odours are produced in cold weather conditions.
- Moisture is required for biological activity to take place. The bacterial activity slows and can be stopped as manure is dried. Moisture also makes anaerobic conditions more likely in the manure and thereby encourages the activity of odour-causing anaerobic bacteria.
- Type of waste material or manure affects the types and quantities of gases produced. For example, liquid poultry manure will produce more hydrogen sulphide than solid manure from broilers. Also, the addition of milkhouse wastes to manure storage can worsen the odour problem.
- Particle size or surface area also affects the rate of odour generation. The greater the surface area present, the faster the bacterial action proceeds.
- Chemicals may alter the process to reduce or increase the number and types of odours produced.

Often, odours are formed as the manure breaks down in storage, and remain trapped in the manure until they are released when the manure is agitated, moved, or spread. The goal of odour management is to reduce the frequency, intensity, duration and offensiveness of odours and to manage the farm in a way that creates a positive attitude toward the operation.

Roof exhaust outlets to maximize air dilution.

BMPs for Odour Control in Livestock Facilities

- Where storage is outside the facility, collect and transfer manure from the barn to storage on a daily basis.
- Ensure that sufficient bedding is added to absorb liquids with solid manure handling systems.
- Maintain water systems to prevent leakage.
- Use a pressure washer to clean buildings.
- Clean and disinfect buildings between successive groups of livestock.

- Keep dust levels low since odours are absorbed and carried in the air on dust particles; add moisture or oil to feed as a dust suppressant.
- Maintain recommended air flow through livestock buildings.
- Clean and maintain ventilation fans and shafts.
- Locate exhaust outlets for maximum air dilution; higher outlets generally provide greater dilution of exhaust gases.
- Locate exhaust outlets to take advantage of the prevailing winds; face them away from the nearest neighbour's residence if possible.
- Do not exceed recommended animal densities for livestock buildings.

BMPs for Odour Control in Manure Storage

Most odour-causing gases are formed when manure is in storage. In practice, most manure storage is anaerobic. The anaerobic conditions promote odour production. These gases either escape from the storage to cause immediate problems or are released later during spreading.

Liquid versus solid- Typically fewer odours are produced by solid manure handling systems than by liquid systems. An undisturbed solid manure stack is self sealing so few odours are given off until the pile is disturbed. With open liquid storage, odours are common. Weather, as well as the addition of manure, can agitate the slurry-causing gases to be given off.

Covering a storage is an effective way to minimize odour generation. Storage covers:

- Reduce occasional manure agitation caused by wind and rain;
- Reduce the movement of odourous air from storage areas to neighbouring residences.

When evaluating manure storage options, consider the following guidelines to reduce the potential for nuisance odours:

- Provide additional storage volume for greater flexibility in the timing of manure application. This can reduce the likelihood of storage overflow and permit application to coincide with the most appropriate timing and weather conditions.

Combination indoor/outdoor feedlot.

- Separate the liquid and solid portions of manure in storage to reduce the promotion of anaerobic conditions.
- Avoid the addition of silage effluent and waste forage products to the manure storage reservoir. These combinations create strong odours.
- Discharge the inlet pipe below the liquid level to avoid surface agitation in a liquid storage system.
- Plant a buffer zone of trees or construct an earthen berm around the manure storage to reduce the movement of air over the manure surface. This has the added benefit of removing the storage from the sight of neighbours.
- Treatment technologies are available and can be used in rare cases when dealing with severe odour problems. Treatment systems must be designed to handle the manure volumes generated by the livestock operation.

Feedlot Management

Feedlots are intensive operations where livestock are kept in a confined area and all food and water are delivered to the animals. The livestock can be either totally confined indoors, outdoors, or a combination of the two. Animal densities for outside lots will depend on whether the lot has a soil base or is hard surfaced. Space requirements can be less than 4.5 square metres (50 sq ft) per head for yearling beef cattle.

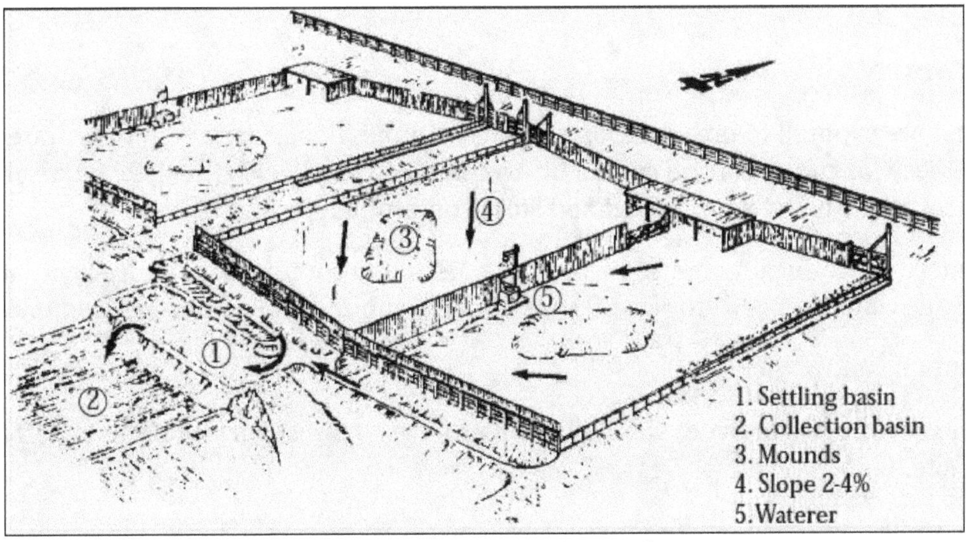

Example of a feedlot area run off control system.

Site Selection

- Avoid sites with permeable soils and fractured bedrock. Groundwater contamination would be a high risk on these sites.
- Maintain recommended property setbacks from watercourses, wells, and neighbouring properties.

- Provide adequate lot slopes on outside lots for surface drainage.
- Allow for potential expansion.
- Prevailing wind direction should be taken into account in siting livestock facilities.
- Wind protection will enhance livestock performance.
- Ensure that upslope runoff is diverted away from the feedlot.

Solid Systems

Most confined livestock areas use a bedded pack. For well-bedded areas much of the liquid is absorbed, resulting in minimal seepage. Regardless of the amount of bedding, however, all seepage and runoff must be contained on the feedlot property.

- A thorough cleaning of the feedlot once a year is recommended. Over cleaning will tend to remove the compacted and impervious soil and increase the possibility of downward nutrient movement.

Liquid Systems

Liquid systems use a slatted floor barn with no bedding. Liquid manure systems for beef operations are uncommon because of the associated high costs of the systems compared to other alternatives.

Runoff Control

Feedlots that are exposed to precipitation are likely sources of surface and groundwater contamination. The need for runoff control cannot be overemphasized. Legislation now requires that feedlot areas incorporate a system to collect and store contaminated runoff.

It may be more economical to house beef cattle in a totally confined, naturally-ventilated facility than to provide collection and storage of contaminated runoff from an outside feedlot.

BMPs for runoff control include:

- Diversion ditches or dykes should be constructed to direct surface water runoff away from the site.
- Grass filter strips where appropriate.

Feedlot runoff will be easier to control if the feedlot yard is graded. This requires a 2-4% slope away from the feed area or shelter. Runoff from each yard should be directed to a collection basin or to the manure storage. The size of collection basins to store runoff from confined livestock areas depends on the size of the runoff area and the amount of precipitation. For PEI, allow 0.61 m^3/m^2 (2 cubic feet per square foot) of surface area. Runoff control systems should be designed by an engineer. The volume of runoff can be reduced by limiting the size of the confinement area. Collected contaminated runoff can be either applied to the land or treated in a constructed wetland.

Odour Management

Frequent cleaning and a high level of sanitation are the most effective ways of minimizing odours from feedlots. Key measures are:

- Keep the animals as clean and dry as possible. Wet manure on the warm body of an animal accelerates bacterial growth and increases odour.
- Scrape manure from the lot surface frequently. Drainage becomes less effective as manure accumulates.
- Provide well-bedded dry resting areas. This results in cleaner cattle, better overall sanitation and less odour from the lot area.
- To avoid continuous wetting of manure, prevent watering facilities from overflowing.
- Time the cleaning process in relation to seasonal weather conditions i.e. temperature, wind, etc.

Overall, maintain a neat appearance around the feedlot. Well-placed visual screens and shelterbelts ensure a positive public perception.

Land Application

Spreading manure on land is a highly desirable method of recycling a natural, organic byproduct of livestock production. A sustainable agricultural system should include manure as a fertilizer for crop production. Manure is readily available with a minimal input of energy and can significantly decrease crop production costs.

Most livestock operations are surrounded by large areas of productive agricultural land. To prevent damage to crops, minimize the risk of pollution and obtain the maximum benefit of the manure as a fertilizer, manure application rates should match the crop nutrient requirements. Too much of a good thing can lead to problems. Manure is an excellent fertilizer which poses an environmental risk only when mismanaged.

Animal manure can be a valuable soil amendment. When properly managed, it not only acts as a source of plant nutrients, but also helps improve soil tilth, structure, aeration and waterholding properties through the addition of organic matter.

Field application of solid manure.

To maximize the utilization of manure nutrients by crops:

- Have a sufficient land base for manure spreading.
- Test soil and manure to determine nutrient levels.
- Understand the release rates for nutrients in manure.
- Calculate crop nutrient demands.
- Prevent the loss of nutrients in surface runoff.
- Reduce the loss of nitrogen to the atmosphere.
- Minimize soil compaction and problems with soil structure.
- Prevent leaching of nitrates into groundwater.
- Prevent pollution of waterways by manure runoff.
- Minimize odours during spreading.

Application Rates

Manure application rates should be determined as part of an overall nutrient management plan. Do not try to provide all nutrients for a crop with manure. It is not likely that manure will release its nutrients at the right balance and time for your crop. Also, not all manure will have the right composition to meet crop requirements.

- Test the macro-nutrient (nitrogen, phosphorus and potassium) content of your manure.
- Test the soil to determine nutrient levels.
- Know the nutrient needs of the crop being grown.
- Set a realistic target of providing up to 75% of the required nitrogen with manure. The balance would be provided by fertilizer. You need to know how much is applied in either case.

While the nitrogen requirement is the key factor in deciding the amount of manure to apply, phosphorus and other elements can also increase to excessive levels in fields where manure is applied every year. An adequate land base is important to get the full benefits of manure. Long-term benefits increase if manure is spread over larger areas. Avoid yearly applications to the same land unless both a soil and manure test show there is no risk of reaching excessive nutrient levels.

The rate at which you can apply liquid manure will also be limited by the soil's ability to soak up the liquid before it runs off. Tillage before application may help if high rates are planned.

Spreading Considerations

- Manure should be incorporated into the soil as soon as possible after spreading. This will minimize the potential for odour complaints and pollution from runoff and will ensure that maximum fertilizer benefits are gained from the manure. It is recommended that surface applied manure be incorporated within 24 hours of application.
- Injecting liquid manure directly into most soils is the best practice if it can be done before preparing the seed bed or during the cropping season. Nutrients are readily available to growing plants.
- To avoid soil compaction problems, do not apply manure under wet soil conditions.
- Manure must not be discharged or allowed to enter any watercourse.
- Manure should not be spread within 30 metres (100 ft) of a watercourse on slopes less than 5% and within 60 metres (200 ft) of a watercourse on slopes greater than 5%.

Timing Considerations

It is usually best to apply manure before, or early in, the growth stage of any crop. Some forms of nitrogen are available immediately to plants. In addition, nutrients in organic form may be released throughout the growing season. If manure is spread late in the growing period or after the growing season, there is reduced benefit to the crop, and there is an increased risk of nitrate leaching to groundwater or surface runoff contaminating watercourses.

Manure Application Calendar

November to mid-April:

- Manure should be going into storage, not on fields.
- Do not spread on frozen, bare, or snow-covered land.

Mid-April to mid-June:

- Apply to land growing annual crops before planting.

Mid-June to August:

- Inject liquid manure between rows of growing row crops.
- Apply manure to cereal land immediately after harvest and prior to conservation tillage.

September to October:

- Apply manure to grassland. Avoid applications in areas subject to concentrated runoff and avoid tillage until after October 15.
- Apply to annual crop lands that will be planted with winter cover crops.

Odour Management

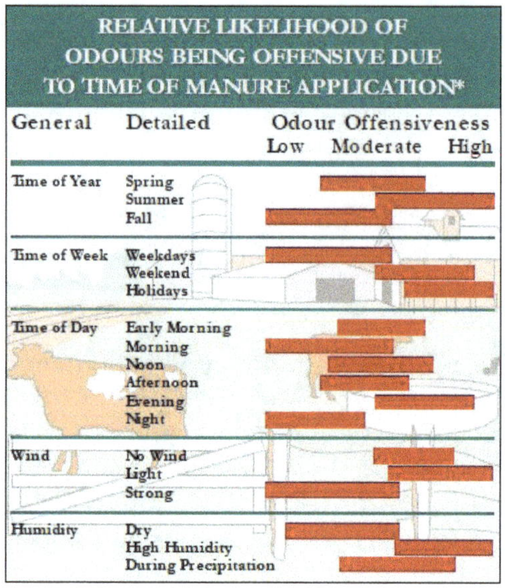

Manure spreading is the most common cause of nuisance odour. There are many factors that contribute to the production of odours during application. The following management practices are the best method of minimizing the possibility of complaints:

- Keep transport equipment clean and well maintained to ensure that manure is not deposited on public roads. Do not overload equipment. If a spill occurs, clean the road promptly.
- Avoid transporting manure on public roads during periods of high traffic such as rush hours before and after work, or during lunch break.
- Check the weather conditions before spreading. The best weather for spreading is sunny days with windy, cloudy nights. Sunshine will dry the manure quickly, preventing further odour production. Turbulent air movement dilutes odours. Rain removes odours from the air. However, the worst conditions are damp, humid weather with light winds. Still air keeps the gases in the area and moist conditions allow for more odour production.
- Spread in the morning when air is warming and rising, rather than late in the afternoon.
- Consider when possible the implications of spreading on holidays and weekends when neighbours are most likely to be affected by odours.
- Notifying neighbours prior to spreading on adjacent properties is a "good neighbour" policy.
- Do not exceed recommended rates of application for your soil type. Generally, the worst period for odours is during the first 12 to 48 hours after spreading. With very heavy application rates, odours could last up to 10 days.
- Incorporate manure into the soil as soon as possible after application. Spread and till methods reduce the release of odours. Injection of manure directly into the soil is an excellent method of odour control. With injection, odours are less detectable at 70 metres (230 ft) than they are at 400 metres (1,300 ft) from surface applied manure.

- Apply composted manure to pastures and hay fields. Where this is not possible, apply manure in a very thin layer so that it will dry in five days or less. This will also prevent fly propagation.
- Keep the discharge height of the slurry as low as possible to reduce odours during land application.
- Choose discharge methods that are most effective for odour control. From most to least effective are: Dribble bars or booms, Bottom discharge tanker, Top discharge tanker

Manure Treatment

Liquid manure injection system.

Treatment of manure is designed to reduce the pollution potential, make handling easier and increase the value of manure. Treatment can be either physical, biological or chemical.

Physical treatments such as separation of solids from liquids are usually considered primary treatment.

Biological treatment methods are usually categorized as either anaerobic or aerobic to describe the type of bacteria that are encouraged to break down the solids in the manure. Composting, naturally aerated lagoons, oxidation ponds, mechanical agitation or pumping and air injection are examples of aerobic treatment methods. Anaerobic methods include anaerobic ponds and digesters.

Many of these treatment systems have not been totally successful in on-farm applications. Either the treated wastes would still pollute the environment, or the systems are too costly to be economically feasible. Producers should give serious consideration to the economic and social benefits derived from treatment technologies before investing. As technologies continue to be developed, a suitable system may be found.

On-farm Composting of Manure

Composting is the aerobic decomposition of organic materials by microorganisms under controlled conditions. During decomposition, the microorganisms consume oxygen while feeding on organic matter. Composting reduces both the volume and mass of the raw materials while transforming them into a valuable soil conditioner.

The benefits of compost:

- Compost adds organic matter, improves soil structure, reduces fertilizer requirements and reduces the potential for soil erosion.
- Composting involves an increase in expenditure, however the increased market potential and soil conditioning properties offer benefits.
- Markets for compost are readily available. Potential buyers include home gardeners, landscapers, vegetable farmers, operators of golf courses, etc.
- Composting reduces the weight and moisture content and increases stability of manure. Compost is easier to handle than manure and stores well without odours or fly problems, thus lowering the risk of pollution and nuisance complaints.
- Composted manure is less susceptible to leaching and further ammonia losses. Composting high-carbon manure/bedding mixtures lowers the carbon/nitrogen ratio to acceptable levels for land application.
- Proper temperatures within the compost pile will reduce pathogens.
- Potential reduction in soil borne plant diseases.

Composting systems used on farms:

- Passive composting - Involves simply stacking the materials in piles to decompose over a long period of time with little agitation and management. Windrow composting - the materials are formed into long narrow windrows which are mechanically turned.
- Aerated static pile - The most common approach, uses blowers to force air though pipes and into the pile.
- In-vessel composting - The materials are contained within bins, reactors, or buildings where a high level of control of moisture and oxygen is provided.

In terms of cost, labour, management and process speed, the windrow and aerated static pile systems are comparable. In-vessel composting is generally more expensive but results in better control over the process, a higher quality product, and less odour.

The location of a composting site should provide:

- Easy access with a minimum of travel and materials handling.
- A firm surface to support vehicles under varying weather conditions.
- Appropriate separation distance from wells, watercourses and neighbours.
- Minimal risk of groundwater contamination.
- Good surface drainage.
- Grading for containment of surface runoff.

Compost applications to land should be based on soil test results and crop needs. This is to prevent a nutrient imbalance from occurring and to make efficient use of compost.

Fly Control

Flies near livestock and poultry facilities and manure storage areas are a nuisance to farm operators and neighbours. Flies may also transmit disease from one farm to another. A successful fly control program can involve:

- Regular removal of manure and wet feed from the building: at least once every seven days during the fly breeding season to break the reproduction cycle.
- Avoiding the scattering of manure and feed outside the building during barn cleaning operations.
- Keeping the manure collection area dark.
- Providing screens on all openings in buildings.
- Keeping the manure in enclosed structures if possible.
- Prompt disposal of dead animals and afterbirth.
- Regular cleaning and disinfection.
- Use of biological controls i.e. parasitic wasps.

In situations where these management options are not sufficient to control the problem, spraying with insecticides may have to be considered.

Milkhouse Waste

Handling milkhouse wastewater has become increasingly important as dairy operations become larger and more automated. Quantities and strength of wastewater from milking parlours vary from farm to farm.

Modern milking parlours and pipeline milking systems utilize large quantities of water. The volume of water used depends on:

- The management practices associated with the milking facility;
- The type of milking system;
- The bulk tank system;
- The floor wash down;
- Other uses ie. udder wash, water conditioners, etc.

The average daily production of washwater on a per milking cow basis is 14.1 litres (3.1 gallons). The disposal of this washwater has become a major environmental concern.

Wastewater contains milk solids, fat, detergents, acid cleaners and sanitizers, manure, soil particles, and other substances.

Environmental Concerns

The following table shows typical waste strengths for milkhouse washwater:

Milking Centre Washwater	
Parameter	Concentration (mg/l)
Total Solids	1417-3506
Suspended Solids	171-996
Oil and Grease	5-330
Biochemical Oxygen Demand	207-1530
Chemical Oxygen Demand	542-4554
Total Phosphorus	35-288
Total Nitrogen	14.9-37.4

Best Management Practices

Proper milkhouse washwater management should consider options for reduction and reuse.

Reduce

- Manual washing and prepping of cows uses less water than automated systems.
- Mechanically removing manure and wasted feed from the parlour prior to wash down reduces waste volume and strength.
- Manually check water hardness and iron content, and calibrate cleaning equipment annually. Adjust chemical cleanser concentrations based on the quality of the washwater.
- Design the milking parlour to minimize washwater requirements. Drain locations and floor slopes are important.

Reuse

- Feed the first rinse of milking equipment to calves. This will reduce the amount of milking centre washwater by 15-20%.
- Feed pre-cooler water to livestock. Pre-coolers are used to lower milk temperature before it enters the bulk tank.
- Many new dairy operations have underground tanks to store washwater. Use this water to wash parlour floors and drain to manure storage.
- Recycling washwater reduces the amount of chemical cleansers required. Washwater can be used from one cleaning cycle to the next.
- Make sure that reused water does not increase bacterial counts.

Handling and Treatment Options

Regardless of the disposal system used, it must be properly designed, installed and operated. To select and design the best system for your farm, you need to know your approximate daily wash-water production. Measure your actual water use by installing a water flow meter, or estimate it using a calibrated pail.

Liquid Manure Storage

Farms equipped to handle liquid manure can divert milkhouse effluent to the liquid manure storage. The combined milkhouse waste and manure is eventually applied to the land with liquid manure following proper manure management guidelines.

Advantages

- An existing storage with adequate capacity can be utilized to store milkhouse waste;
- Better agitation can be achieved through the extra volume of liquid milkhouse waste making it easier to pump to a spreader;
- Storage can be sized to handle the additional wastewater;
- Best option if washwater contains the first rinse and has a high solids content.

Disadvantages

- Lowers fertility value of manure;
- Requires up to 25% increase in manure storage capacity;
- The increased volume must be spread on fields, adding to handling costs.

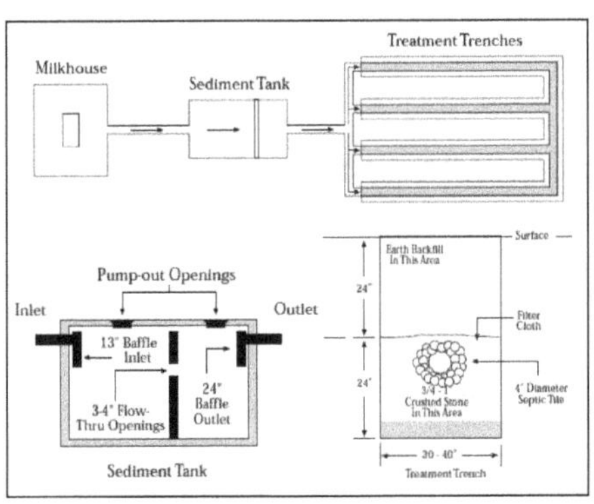

Sample Illustration of Sediment Trench & Treatment Trenches.

Settling Tank and in Ground Disposal Field

This system is similar to a household septic system but instead handles the wastewater for the milking centre. It also requires proper site conditions. If the soil is too shallow to bedrock, has low permeability or a high water table, another management system may be required. This system requires careful design installation and management to ensure long-term success.

Advantages

- Does not impact the capacity of manure storage and handling systems;
- Relatively low cost.

Disadvantages

- First rinse, milk from treated cows or colostrum must be diverted from the system;
- Does not work in areas with low permeability or high water tables;
- Requires careful management and maintenance;
- Sludge levels within the tank should be checked and removed as required.

Conventional in ground disposal fields have been used extensively in the past and with proper design, management and soil conditions, these systems can work quite well.

Historically, many systems have failed due to the disposal of excessively high strength milkhouse washwater through the system. You may consider this system if:

- You do not have or are not planning to build a liquid manure or runoff storage;
- The soil has good drainage characteristics;
- Your washwater is relatively free from solids;
- You are willing to restrict the water that enters the system to a minimum;
- You are willing to collect the first rinse from the milking equipment;
- You are willing to prevent milk from going down the floor drain.

To prevent whole milk from reaching the disposal field system, it is recommended that a double compartment septic tank with proper baffles on the inlet and outlet pipes and a storage capacity for several days of milkhouse effluent be installed. Allowing the milk several days residence in the tank will allow most of the milk fats to separate and form a layer at the top of the tank which must be periodically pumped. A minimum retention time of four days is recommended, but six days is preferred.

Research and experience have shown that constructing the disposal fields to promote aerobic conditions for a period of time between milkings will greatly improve the reliability and useful life of these systems. This will allow microbial activity to oxidize some of these materials and prevent trench sealing.

Vegetative Filter Strip Bed

Testing in the Maritimes has shown that a grassed filter strip can also be an effective and economical milkhouse wastewater treatment alternative. As the wastewater flows down a filter strip it evaporates or infiltrates into the soil. The soil and plant media filter out and biodegrade the fine solids and organic material. Nitrogen and phosphorus are taken up by the plant life and absorbed to soil particles.

Vegetative filter strip bed.

The design of an effective filter strip includes a settling basin ahead of the filter strip, a spreading device at the entrance of the filter strip to ensure even flow across the strip and provisions to alternately apply wastewater to two parallel strips. This will allow the filter strip a rest period during which no wastewater is applied.

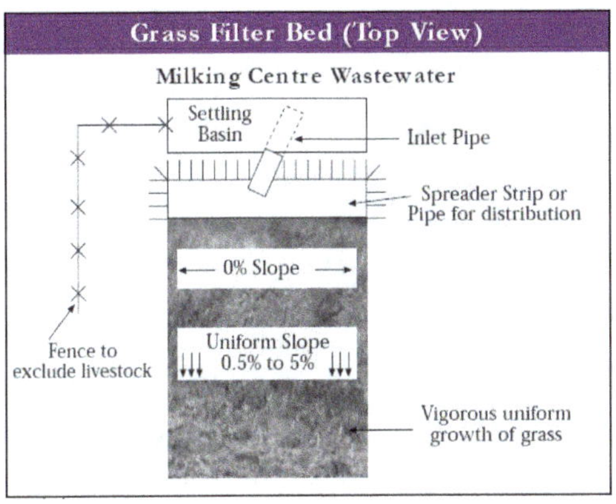

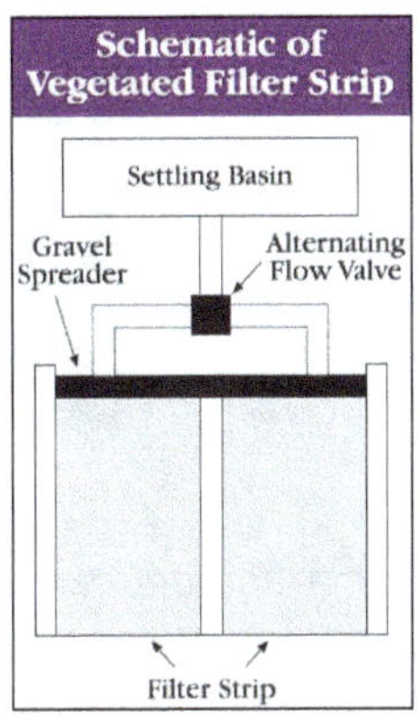

Advantages

- Does not impact upon existing manure storage or handling;
- Relatively low cost.

Disadvantages

- First rinse, milk from treated cows or colostrum must be diverted from the system;
- Requires careful management and maintenance;
- Solids must be removed regularly from the settling basin to prevent overflow to the grass filter strip;
- May not provide optimal treatment during the winter months.

Constructed Wetlands

Constructed wetlands are shallow, man-made aquatic systems that can provide an environment for treating agricultural runoff and wastewater. Constructed wetlands have been utilized as treatment systems for a number of wastewater sources including:

- Milkhouse washwater;
- Manure storage and feedlot runoff;
- Drainage tile outflow;
- Agricultural field surface runoff; and
- Food processing wastewater.

Constructed wetlands utilize a series of physical, biological and chemical processes which facilitate the treatment of wastewater.

Constructed wetland.

Wetlands have been constructed on agricultural operations throughout Atlantic Canada and many have been extensively monitored. The concentration of waste-water pollutants including suspended solids, nitrogen, phosphorous and faecal coliforms, as well as BOD and COD levels, have been reduced by 70-98 %.

Design Considerations

- Before construction, hire a qualified engineer to design the wetland and obtain required building and environmental permits from the PEI Department of Fisheries, Aquaculture and Environment.
- Constructed wetlands should only be designed as secondary or tertiary wastewater treatment systems. The size of a wetland must be based on the inflow volume, the concentration of pollutants in the wastewater, and the desired level of treatment.
- Site selection is important. The wetland will be more economical to construct if it is located close to the wastewater source and if the wastewater can flow by gravity to the wetland.
- Perform soil permeability tests early in the planning stage. If the hydraulic conductivity is greater than 1×10^{-5} cm sec^{-1}, a clay or synthetic liner will be required. This will greatly add to the cost of the wetland and may be a reason to consider other options.

- Many agricultural wastewater sources produce small volumes of effluent. During the summer, evaporation rates from the wetland are often higher than inflow volumes. Additional water from other sources such as roof gutters may have to be added to the system.
- Wastewater must be retained in a settling pond prior to entering the wetland to allow for adequate separation of solids. This pond should be less than 1 metre (3 ft) deep to reduce odour potential.
- Wetlands are more efficient during summer months. It may be desirable to design the settling pond to be large enough to store the entire volume of wastewater produced during the winter and to discharge it to the wetland during the summer.
- The proposed site must be surveyed to produce an accurate topographical map.
- Constructed wetlands may contain one cell or several individual cells depending upon the topography. If the construction site is on a slope, it may be desirable to construct individual cells in a terrace type system. The length of each cell should be twice the width. The topography should be relatively level over the entire cell to ensure an even depth of water. Individual wetland cells should include both deep and shallow zones. Shallow zones should have water depths ranging between 15-30 cm (6-12 in). Deep zones help to evenly distribute water and add to the retention time as wastewater passes through the wetland. Deep zones should constitute 25% of the surface area and they should be at least 1 metre (3 ft) deep to prevent growth of aquatic plants.
- The outflow from a constructed wetland should receive final polishing by discharging it to a tertiary pond or grassed waterway.
- Aquatic vegetation (cattails and bulrushes) is best established by transplanting root stock from a nearby natural wetland. The roots should be planted at a density of at least one plant per square metre (10 sq. ft). A permit is required from the PEI Department of Fisheries, Aquaculture and Environment to remove cattails from existing wetlands.

Management to achieve maximum performance of the wetland may include:

- Eventual removal of solids from the pretreatment settling pond.
- Possible addition of water to the wetland during prolonged dry periods to prevent stress of aquatic vegetation and cracking of a clay liner.
- Eventual excavation of sediment and plant material from the wetland cell if there is an accumulation of phosphorous.
- Control of muskrat populations.

Flocculator

Another innovative new technology recently introduced to the Maritimes involves the use of chemical treatment reactors or flocculators to remove the majority of phosphates and suspended solids from the milkhouse effluent.

The effluent is transferred from the milkhouse to a reactor and a proportionate amount of hydrated lime is added to the reactor. The mixture is allowed to settle undisturbed for two hours. Then the clarified liquid is discharged to the disposal field system while the sludge is sent to the manure storage. This technology is relatively new and its cost efficiency for treatment of milkhouse waste requires more evaluation.

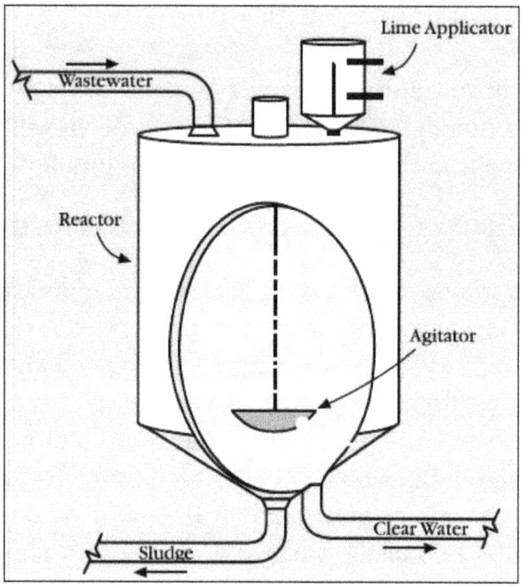

Milkhouse wastewater flocculator.

Waste Forage

Waste forage is a normal by-product of any livestock feeding system. It includes waste hay, waste silage and silage effluent.

Hay

When hay is harvested, it is allowed to dry to less than 15% moisture. When it is stored and fed, it poses no environmental hazard. Hay that is wasted during feeding usually becomes incorporated with the straw and manure.

Silage

The potential for waste from silage feeding systems also exists. However, because silage is harvested and stored at a much higher moisture content than hay, there is also the potential for leachate or seepage waste from storage areas. Silage can be stored in vertical silos, horizontal silos, or in round bales wrapped in plastic.

Environmental Concerns

While forage waste is costly to producers, it is also an environmental concern.

- Silage seepage can leach into the groundwater and can contaminate watercourses if runoff is not controlled.

- Burning waste forage can cause nuisance odours and contribute to air pollution.
- Silage waste can also create nuisance odours.

Waste forage being used as a mulch for erosion control after potato harvesting.

Best Management Practices

Reduce waste forage by:

- Storing bales of hay under cover.
- Harvest and store only as much forage as will be required for the coming year.
- Harvest silage at the optimum moisture content to minimize the potential for seepage.
- Use waste forage as a mulch to provide protection from soil erosion in recently harvested potato fields rather than hauling it to the woods or burning it. It will add organic matter to cultivated soils.
- Compost waste hay and silage. This will require addition of a nitrogen source such as manure.

Seepage from Farm Silos

Silage seepage presents several concerns for the agricultural industry.

- Contamination of surface and groundwater.
- Deterioration of the silo structure.
- Odours.

Silage effluent has extremely high BOD values, which are approximately 200 times as strong as raw domestic sewage. A significant discharge of effluent into a watercourse can remove so much oxygen that fish and other aquatic creatures die immediately.

Harvesting silage at proper moisture levels will minimize the risk of seepage.

With respect to groundwater quality, silage leachate contains nutrients, acids, minerals and bacteria. Nitrate-nitrogen is the most significant groundwater contaminant from this group. The greatest percentage of silage seepage is produced within 5 to 10 days after filling the silo.

Best Management Practices

- Seepage from the silo, along with the surface water runoff from open bunker silos, should be collected and stored since this material is highly contaminated. During the cropping season this contaminated material can be spread regularly on land.
- Harvest silage/haylage at low moisture, i.e. below 60% moisture content for tower silos and below 65% for horizontal silos.
- Adding absorbents which are designed to take up excess moisture will result in low or no seepage production. Material that can be used include oatmeal, dried sugar beet pulp and dried corn cobs. To be effective, enough material must be added to absorb the anticipated seepage.
- Silos should be covered - this prevents rain water from entering and leaching through the silage/haylage.
- Divert all surface water away from the silo site.
- For new silos, install seepage collection and storage systems.

Dead Stock Disposal

Composting of dead stock carcasses.

The disposal of dead animals must be in accordance with the PEI Guidelines for Disposal of Dead Farm Livestock. These guidelines are designed to protect the public and animal health and to reduce the risk of contamination of drinking water supplies and surface water resources. All livestock and poultry operations need a management plan for dead animals. Afterbirth in livestock operations and eggs in poultry operations are additional wastes which should be handled in a similar manner.

Environmental Concerns

Dead animals are a health risk to humans and farm animals. They may be carriers of disease. If not promptly removed or disposed of, carcasses will also attract rodents, flies, and nuisance wildlife, as well as producing offensive odours.

Options for Disposal

Dead animals should be removed from the building as soon as possible and disposed of in an approved manner within one day. Livestock and poultry cannot be disposed of in manure storage or be spread onto the land with manure. Under no circumstances should dead animals be left lying around the farm buildings for an extended period of time. Feeding carcasses to wildlife is not an acceptable disposal method and should not be used.

While health and environmental concerns are most important, ensuring that disposal activities are not unsightly will create good will in the community. Trees, shrubs and windbreaks can be used to screen disposal sites. Disposal options are covered in order from most to least acceptable.

- Dead Animal Removal Service

A dead animal and poultry removal service removes carcasses to be rendered. Poultry and other small animals must be stored in freezers and held for pickup. Livestock should be picked up within 48 hours of death. For a current list of livestock removal companies, call the PEI Department of Agriculture and Forestry.

- Composting

Composting of poultry and other small agricultural animals is considered a low cost, environmentally acceptable method of disposal of dead stock. The use of composting technology in the disposal of poultry and hogs is becoming more widespread.

Some basic requirements for carcass composting are:

- An aerobic environment must be maintained.
- A carbon source will be required to ensure that the proper carbon:nitrogen ratio of 25:1 is present for the composting process to take place. Sawdust is an ideal carbon source.
- Temperatures throughout the compost pile must exceed 55 °C (130 °F) for adequate reduction of pathogen levels.
- Moisture content of the compost pile should be in the 50- 60% range.
- The composter should be located in an area that is well drained, accessible and away from

areas that are sensitive to groundwater contamination. If a facility is constructed for composting, it should consist of a concrete pad, roof, and rotresistant construction materials.

- The composting site must be located a minimum of 90 metres (300 ft) from a watercourse or domestic well.
- All contaminated runoff from the compost site must be collected. Clean surface water should be directed away from the composting facility.
- The finished compost can be spread on the land.
- Capacity of the composting facility must be sufficient to dispose of the normal mortality rate. Expanded capacity to contain an excessive mortality rate is desirable but not essential.

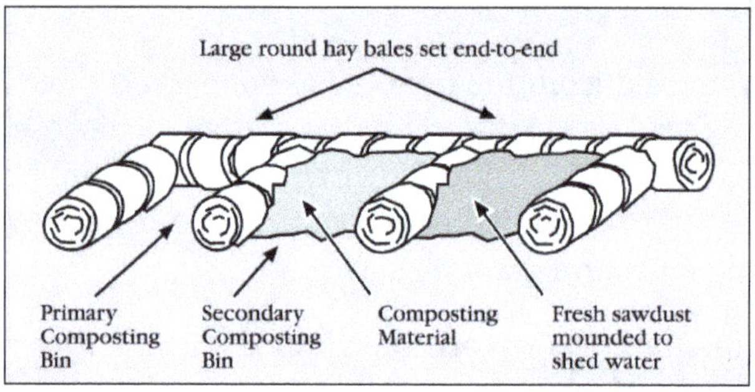

- Subsurface Burial

Burial should be considered only as a last resort. Subsurface burial is not recommended due to the potential for groundwater pollution. It is permitted as long as the criteria listed below are met.

Roofed composting facility.

Burial sites should be staggered throughout the property, not concentrated in one location. Burial sites must be located:

- At least 300 metres (1,000 ft) from any drinking water supply or well. With prior written approval from the Department of Fisheries, Aquaculture and Environment, a burial site may be located closer than 300 metres (1,000 ft), but at no time will the Department approve a site less than 150 metres (500 ft) from a drinking water source.
- At least 60 metres (200 ft) from any fresh water stream, pond, estuary or coastal area.
- At least 30 metres (100 ft) from any public right of way.

All buried poultry and livestock must be covered on the same day they are buried with a minimum of 0.6 metres (2 ft) of earth. Subsurface burial should only be considered under the following conditions.

- At locations where the water table does not come within 600 mm (24 in) of the pit bottom and where soil is well aerated.
- At locations where the floor of the burial pit is at least 0.6 metres (2 ft) above bedrock.
- At locations not subject to surface runoff, ponding or flooding.
- Open trench pits are not acceptable.

ORGANIC WASTE MANAGEMENT

Organic Waste Management (OWM) operates an organic waste removal service that collects "source-separated" food scraps from your premises and disposes of them in an environmentally sustainable method.

Composting is a managed process which utilizes microorganisms naturally present in organic matter and soil to decompose organic material. These microorganisms require basic nutrients, oxygen, and water in order for decomposition to occur at an accelerated pace. The end-product, compost, is a dark brown, humus-like material which can be easily and safely handled, stored, and used as a valuable soil conditioner. The composting process is dependent upon several factors, including: the population of microorganisms, carbon to nitrogen ratio, oxygen level, temperature, moisture, surface area, pH, and time.

Aerobic Composting

The composting process involves microorganisms feeding on organic material and consuming oxygen. The process generates heat, drives off moisture, and reduces bulky organic waste into a beneficial soil-like material containing nutrients, humus and microorganisms in just a few months. Material in an unmanaged pile of organic debris will eventually break down but the process will take a long time and may result in odor or other nuisance problems due to poor aeration.

What to Compost

Most organic waste materials generated by a greenhouse can be composted. Large material will need to be shredded before it is added to a carefully-constructed compost pile. Some material may begin to decompose in a storage pile but full composting will not occur until the material is mixed and managed in the correct proportions of carbon to nitrogen (C:N ratio), with adequate airflow and moisture.

Composting is an excellent method of recycling grass clippings. However, do not compost grass clippings or any other plant residues that have been treated with herbicides. If carried out properly, it can reduce the potential weed seeds and diseases from being reintroduced into the fields. The finished compost is a stable organic material which is a useful soil conditioner or nutrient source. Due to the characteristics of fresh grass clippings (high-moisture, high-nitrogen content and small particle size), co-composting with a high-carbon bulking agent is essential.

Unacceptable materials for composting: Chemically treated wood products, plastic (e.g. pots, bags and sheet film), unprocessed sod and chunks of soil, large bulky idems (e.g. stumps, pallets, concrete and asphalt).

Acceptable Materials for composting: Green and woody plant clippings and trimmings, soil, plant media, untreated wood and uncoated paper scraps.

Site Selection for Compost Piles

Proper site selection is a prerequisite to the establishment of safe and effective composting operations. The location of a composting operation directly impacts the amount of site preparation required and the measures needed to satisfy environmental and regulatory requirements.

Basic Composting Process

The general steps in the biological process which creates compost are the same regardless of the raw materials being composted or the size and complexity of the production facility. A compost must pass through all of the steps outlined here in order for it to be considered of high enough quality for use in organic potting mixes.

The progress of organic matter decomposition during composting can be followed by monitoring the temperature of the compost pile. During the initial phase of composting the temperature of the pile increases rapidly as the population and activity of decay microorganisms increases in response to the readily decomposable carbon in the raw materials. The goals are to reach a temperature between 131°F or more and to maintain this temperature range until the microorganisms begin to exhaust the readily available carbon. During composting the pile is turned and remixed several times to ensure complete heating and decomposition.

To comply with the National Organic Program standards compost piles must maintain 131-170 °F for at least 3 days (static pile) or at least 15 days (windrow, turned at least 5 times). High temperatures are necessary to kill any human pathogens especially if farm manure is a component. Also, weed seeds and plant diseases are most successfully killed at high temperatures. Most weed seeds are destroyed at 145 °F.

Following the high temperature phase there is an extended period of gradual temperature decline until the pile reaches ambient air temperature. Now, if the pile is turned, reheating will not occur. At this point the compost is said to be "near maturity", but to ensure that the compost is stable and ready to use, most producers allow some extra time for the compost to "cure". How long composting lasts varies with the method. It could take about 1-2 years in a static unturned pile, 6-9 months if the pile is turned occasionally, or only 1-4 months the pile is turned frequently.

Many types of raw materials can be used for making compost; some common materials are listed in the following table.

Compost Material	Comments
Farm animal manure	Must be composted
Straw and bedding	–
Crop residues	Must be pesticide free
Fruit and vegetable wastes	Must be pesticide free
Food processing wastes	–
Seafood processing waste	–
Grass clippings	Must be pesticide free
Sawdust and other wood wastes	Use in moderation, low nutrient value.
Newspaper	Black ink only, shredded, <25%
Leaves	Shredded

It is important that the raw materials be properly prepared prior to mixing and the start of composting. Most organic materials must be shredded or ground to reduce particle size and help make them less resistant to decay.

During composting, oxygen and moisture levels are critical factors in determining the degree of decomposition which takes place and the length of time it takes to reach a stable product. Oxygen levels below 5% and moisture levels above or below the range of 40-65% inhibit the composting process. Most composting operations aerate the piles (turn) and irrigate them if conditions favor excessive drying. Too little moisture will inhibit microbial activity and slow down the composting process, while too much moisture will restrict the flow of oxygen because all pore space is taken up by water instead of air, and anaerobic conditions will begin to develop. The volume of the finished compost is smaller than the volume of raw materials because of the breakdown of organic matter and the evaporation of water.

When is the compost ready for use? Currently there is no single widely accepted criterion to determine when compost is "done". Measurements of temperature, respiration, ammonia production, pH, and carbon to nitrogen ratio (C:N) are among the potential indicators of compost maturity, but no one factor is completely reliable. Generally, at the end of active composting (heating period) producers allow a "curing" period of about 1-2 months to make sure the compost is stable before it's used.

Checklist of Organic Waste Management

- Most organic waste materials generated by a greenhouse can be composted.
- Avoid composting grass clippings that has been treated with herbicides.
- Compost piles should always be distant and downwind from sensitive neighbors and not sited close to residential property.
- Piles should be protected from surface water and storm water runoff.
- Proportions of carbon to nitrogen are critical to successful composting. The materials being composted will determine the exact recipe for any given operation. Materials with high

carbon to nitrogen ratios, such as 100:1, should be balanced with materials having low carbon to nitrogen ratios, e.g. 15:1.

- Regular turning of the pile will mix the nutrients and re-establish pile structure.
- Moisture content can be adjusted during turning.
- After the compost has gone through several heating and cooling cycles and the original waste has decomposed, the compost process should slowly finish in a curing pile.
- Activities that tend to release odors should be scheduled to minimize negative impacts.
- Consider wind conditions before opening compost piles. Stronger winds can disperse odors but also create dust concerns. Care should be taken to control dust when grinding and turning piles.
- Most odor problems can be avoided, controlled, or minimized by keeping the compost pile aerobic, porous, well aerated, and well mixed. Odor problems are most likely when anaerobic decomposition is occurring.

SLAUGHTERHOUSE WASTE MANAGEMENT

Composting

Composting is an aerobic process by which organic materials are degraded through the activities of successive groups of microorganisms. Substrates vary, and can include different types of organic and inorganic wastes, sewage sludge, pig, cattle or poultry manure, garden waste and municipal solid waste. The positive effects of composts on arable soil have been reported by many authors.

An efficient and satisfactory composting process is dependent on the presence of a high microbial diversity. Different microbial communities predominate during the four consecutive phases involving mesophilic, thermotolerant, and thermophilic aerobic microorganisms. Continual change in environmental conditions in composts (temperature, pH, aeration, moisture, availability of substrates) results in stages of exponential growth and stationary phases for different organisms. Knowledge on the diversity of the microbial communities of compost is increasing rapidly through the use of molecular biology tools, such as the COMPOCHIP microarray.

Among the concerns regarding the composting process and the use of composts in agriculture and horticulture, however, is the survival and spread of animal, human and plant pathogens. Thus, any composting process must be capable of the elimination of any health risk that may be present in the end product. An improperly managed composting process can induce the proliferation and dispersion of potentially pathogenic and allergenic thermotolerant/thermophilic fungi and bacteria.

Two types of composting systems exist: windrow and in-vessel systems. In-vessel composting has several advantages over windrow composting, namely, that it requires less space, provides better control than windrows, and offers a high process efficiency. The higher temperatures achieved however in windrow composting have been found to result in greater bacterial and pathogen reductions. According to Noble and Roberts, the cool zones in in-vessel composting systems where there is no or little turning, are of concern for pathogen survival. Thus, windrow composting of slaughterhouse wastes is the more appropriate approach.

Composting provides an inexpensive alternative and environmentally acceptable method for the disposal of slaughterhouse wastes. The temperatures reached during the decomposition process can kill or greatly reduce most pathogens, thus reducing the chance to spread disease. Although the heat generated during composting results in a reduction of microorganism and pathogen numbers, it is not sufficient to completely sterilize the end product, thus leaving some potential for survival and regrowth of pathogens. The levels of pathogenic bacteria remaining at the end of the composting process are dependent on the temperatures reached and the time the temperature was maintained. Legal requirements in various countries ask for a minimum temperature of 65 °C for at least 6 consecutive days or two 3-day periods >65 °C. Achieving an average temperature of 55–60 °C for 1–2 days is generally sufficient to reduce pathogenic viruses, bacteria, protozoa, and helminth ova to an acceptably low level. However, the endospores produced by spore-forming bacteria such as Clostridium and Bacillus would not be inactivated under these conditions. Typically, pH values exceeding 8 are also obtained in the composting process which also contributes to the inactivation of pathogens. Pathogens in compost can also be inactivated by the action of antibiotics produced by various microorganisms present in the compost.

Composting can potentially serve as an acceptable method of disposal of slaughterhouse wastes. However, some additional treatments would be necessary to inactivate some pathogens in the end product. Composting can also serve as a method of post- treatment of AD wastes. This post-treatment helps to ensure a more complete breakdown of organic matter, and enriches the compost in terms of the nutrients. Composting of the anaerobic digestate would in addition act to reduce pathogen levels in such products.

AH

AH represents a relatively new technology for the disposal of animal carcass material, and other infectious wastes. The process uses sodium or potassium hydroxide to catalyze the hydrolysis of biological materials (proteins, nucleic acids, carbohydrates, lipids, etc.) into a sterile aqueous solution consisting of small peptides, amino acids, sugars, and soaps. In order to accelerate the hydrolysis, the process is usually conducted with pressure and temperature. For the inactivation of microbial pathogens, carcasses must be heated to 100 °C and pressurized at 103 kpa for 3 h. To destroy prion containing material, carcasses must be heated to 150 °C and pressurized at 486 kpa for 6–8 h.

Studies investigating the inactivation of pathogens using AH are limited, however, those that exist, indicate the method to be highly successful. Neyens et al. reported that an AH process conducted at 100 °C, pH ≈10, 120 kpa and for 60 min was able to kill all pathogens. Unfortunately, it was not reported which pathogens were investigated. In a recent study, Dixon et al. investigated the use of AH at ambient temperature for the inactivation of the fish pathogens infectious salmon anaemia virus (ISAV) and Lactococcus garvieae. Both pathogens were inactivated after 48 h of treatment with 1 M NaOH at pH >13 and at room temperature. The efficacy of AH as an alternative for the treatment and disposal of infectious waste was also evaluated by testing for the destruction of specific microorganisms in a study by Kaye et al. AH was found to destroy all representative classes of potentially infectious agents.

The sterile end product of AH can be released into a sanitary sewer, in accordance with local and federal guidelines regarding pH and temperature. Studies conducted by the same authors have shown that by bubbling carbon dioxide into the hydrolysate at the end of the digestion, the pH of the end product can be reduced into the range of pH 8 or less. Considering that the hydrolysate is comprised of a mixture of single amino acids, small peptides, and fatty acids, all excellent growth nutrients for microorganisms, alternative disposal methods for this waste product should be sought. One possibility is its use as a fertilizer, however, another possibility that would require further research, but which would serve in renewable energy production, would be its use in AD. Coupled with AD, an AH unit could produce significant amounts of energy, and at the same time, safely dispose of potentially contaminated slaughterhouse waste products.

An alkaline pretreatment with NaOH was used in a study to enhance biogas production from the AD of corn stover. Different concentrations of NaOH (1–7.5% w/w) were tested and the preparations were subjected to AD. The highest biogas yields were obtained with 5% NaOH. This study shows that AH of slaughterhouse wastes followed by AD should be considered a potential route of treatment.

There are many advantages of using AH for the inactivation of disease agents in slaughterhouse wastes. These include the combination of a sterilization and digestion step into one operation, the reduction of waste volume and weight by as much as 97%, the total destruction of pathogens including prions such as BSE and the lower emission of odors or public nuisances.

AD

AD is a biological process by which organic wastes are decomposed in the absence of oxygen, producing a sludge of agricultural value, as well as biogas, which can be used to generate energy. In recent years, AD technology has been applied and used commercially in the treatment of farm, industrial, and municipal wastes. AD also represents an alternative method for the disposal of animal carcasses and slaughterhouse wastes, with the dual benefits of both eliminating waste material and producing energy, thus saving carbon credits. Concerns exist however as to whether an AD process can successfully inactivate pathogens. Often, a pasteurization step is used to help inactivate pathogens prior to AD. It is also common to include a secondary heat treatment process (e.g. composting or pasteurization) and a minimum storage period at the end of the process for the digestate as an additional measure. Swedish law for example, requires biogas plants that use animal waste to pasteurize the incoming substrate at 70 °C for 60 min prior to digestion, to ensure a hygienically acceptable end product.

The anaerobic process itself requires specific environmental conditions and requires that the bacteria and archaea involved co-operate in a close and efficient syntrophism. Digestion occurs in four major stages (hydrolysis, acidogenesis, acetogenesis, and methanogenesis) and complex polymers are degraded in a stepwise manner to yield CO_2 and CH_4.

AD can be run either under mesophilic (35 °C) or thermophilic conditions (55 °C), and the duration of the process and the effectiveness in destroying pathogens thus varies. A mesophilic process will typically run for 15–30 days, while a thermophilic process will run for 12–14 days. Mesophilic AD processes are reportedly more robust and less sensitive to changes in process parameters than are thermophilic processes, however, gas production is lower and it is more likely that the end product poses a greater pathogen risk if applied directly to the field. Methane producing microorganisms are more sensitive to temperature alterations than other bacteria, and temperature variations as low as 2 °C can reportedly have adverse effects on mesophilic AD processes.

Thermophilic AD on the other hand offers higher methane production, faster throughput, and a better inactivation of pathogens. Despite its advantages, the thermophilic digestion process needs more expensive technology, and requires a higher degree of operation and monitoring. Also, changes of only 0.5 °C can affect thermophilic AD processes.

Monteith et al. seeded a bovine enterovirus and parvovirus into liquid cattle manure, and noted that both viruses were rapidly inactivated by a thermophilic AD process, and not detected after 30 min. Under mesophilic conditions, however, the viruses were able to survive for 13 and 8 days, respectively. Spillman et al. also found that the rotavirus and coxsackievirus B5 animal viruses were rapidly inactivated by thermophilic AD. These findings indicate that viruses are inactivated during AD, but that the rate of inactivation is dependent on the virus, the temperature and the duration of digestion. Even when using a thermophilic AD process, it is advisable to use an additional heat treatment at the end of the process to fully inactivate pathogens capable of surviving AD (i.e. spore-formers). In fact, the European Commission Regulations require a 70 °C/60 min pasteurization step of animal waste prior to landspreading of sludges. It should also be remembered however, that an additional heat treatment would still most likely not allow the inactivation of prions such as BSE.

The high lipid and protein contents of slaughterhouse waste products can cause inhibition of AD processes, as a result of an accumulation of ammonia produced from the degradation of proteins and long chain fatty acid accumulation from the breakdown of lipids. Also, lipids can form floating aggregates and foam that can cause stratification problems due to the adsorption of lipids into the biomass. However, if properly managed, AD can be conducted successfully on slaughterhouse wastes with high methane production, by allowing a progressive acclimatization of the microbial communities to an ammonia rich medium.

Pathogens

In this topic, information on the survival and inactivation of a number of specific pathogens or indicator organisms is summarized.

Escherichia Coli

Escherichia coli is a Gram negative, rod-shaped bacterium that is often found in the lower intestine of warm-blooded organisms. Most strains are harmless, although there are serotypes that cause food poisoning in humans, and strains such as O157:H7, which can cause serious illness or death in the elderly, the very young and the immuno-compromised.

A pasteurization experiment was conducted by Sahlström et al. to determine the survival rates of E. coli O157 after heat treatment. In the experiment, E. coli O157 was found to survive after 30 min at 55 °C, but not after 60 min at 55 °C. Similarly, a treatment at 70 °C for 30 min was sufficient to inactivate E. coli. Inactivation of non-pathogenic E. coli and pathogenic E. coli O157:H7 has been reported during composting of several types of waste, including animal manure and sewage sludge. The persistence however of non-pathogenic E. coli during composting and the finding of non-pathogenic E. coli in mature composts have also been reported. It is possible that in these cases the composting process employed was not conducted for a long enough time at a high enough temperature.

Laboratory scale experiments conducted by Jiang et al. (2003) have shown that E. coli 0157:H7 in bovine fecal matter are inactivated at thermophilic composting temperatures. In their study, Jiang et al. showed that large populations (104–107 CFU/g) of E. coli O157:H7 survived for 36 days during composting in a bioreactor at an external temperature of 21 °C but were undetectable after 7–14 days of composting in the bioreactor at an external temperature of 50 °C. In another study, Lung et al. investigated E. coli O157:H7 survival after cow manure composting. E. coli O157:H7 was not detected after 72 h of composting at 45 °C. Thus, a composting process conducted at a minimum temperature of 50 °C and lasting longer than a week was assumed to be able to successfully inactivate E. coli O157:H7 in manure, and could be assumed to inactivate other E. coli strains in slaughterhouse waste products.

The reductions of pathogenic and indicator bacteria (including E. coli) in animal slurry subjected to mesophilic (35 °C) and thermophilic (53 °C) AD was investigated in a study using both small-scale and full-scale reactors. At small-scale digestion at 35 °C, a T90 value (time required for a 90% reduction in number of bacteria) of 1·8 days was measured for E. coli. At 53 °C, the T90 value was reduced to 0·4 h for E. coli. Similar values were obtained at full-scale. Thus, we would hypothesize that a combination of a pasteurization step prior to AD, followed by an AD process should totally inactivate all E. coli cells present in any slaughterhouse waste material.

No information could be found concerning the use of AH to inactivate E. coli. However, considering that AH was found in a study by Kaye et al. to completely destroy all representative classes of potentially infectious agents as well as disposing of animal carcasses by solubilization and digestion, it can be assumed the process would also totally inactivate E. coli.

Salmonella

Salmonella are ubiquitous and cause illnesses such as typhoid and paratyphoid fever, and very frequently, food poisoning. All serovars of Salmonella are potentially pathogenic to both humans and animals, and the organism is spread in the environment by animal slurry, and sewage sludge. Salmonella is often used as an indicator organism to test whether a particular treatment process has been successful in the inactivation of microorganisms. Many studies have been conducted investigating the survival of Salmonella at various temperatures, for different periods of time. Generally, Salmonella is not able to survive at temperatures above 70 °C. With the exception of S. senftenberg, Salmonellae are destroyed at 56 °C for 10–20 min, although there is better heat tolerance at low water activity and in high fat foods. In a study by Paluszak et al., S. senftenberg W775 introduced into meat material was found to be eliminated by a composting process within 30 h at 68 °C. However, Salmonella has been reported to be able to re-colonize composts when temperatures are reduced near the end of the composting process if the process did not reach sufficiently high temperatures, or if the pile was not adequately aerated or turned.

Temperature is the most important factor affecting the survival of a pathogenic microorganism during AD. Studies by Plym-Forshell showed that Salmonella could be inactivated within 24 h in a thermophilic digestion process, while weeks-months were required in a mesophilic digestion process. However, in a study conducted by Gadre et al., all Salmonella species investigated were inactivated after 10 days at 37 °C. Pasteurization is frequently used prior to AD in order to reduce pathogen numbers. As it is known that Salmonella in sludge cannot withstand more than 5 min of heating at 70 °C, a pasteurization step should act as extra insurance for Salmonella inactivation from slaughterhouse wastes and in animal material after AD.

Literature does not reveal much information on the use of AH to inactivate Salmonella. A study by Mastroeni et al. found AH of crude whole Salmonella extracts reduced the lethal toxicity of the extracts on mice. Extracts were however only incubated with NaOH at a final concentration of 0.25 M for 3 h at 37 °C. This is much lower a temperature than what is used in AH units. Considering this, and that AH was found in a study by Kaye et al. to completely destroy all representative classes of potentially infectious agents as well as disposing of animal carcasses by solubilization and digestion, it can be assumed that the process would also totally inactivate Salmonella species.

Clostridium: Botulism, Black Quarter and Tetanus

Clostridium botulinum causes the disease botulism. The disease can occur following: ingestion of contaminated food, from colonization of the infant gastrointestinal tract, or from a wound infection. Vegetative cells and spores of C. botulinum can be found in soil, decaying vegetation and manure. The disease results from the production of a potent neurotoxin which is released when spores germinate.

Black quarter is an infectious and often lethal bacterial disease of sheep and cattle caused by Clostridium chauvoei. The bacteria can be found in soil and faeces, and once a pasture has become contaminated, the disease usually occurs annually in susceptible animals that graze on the pasture.

Clostridium tetani is the causal agent of tetanus, a disease characterized by muscular spasms that can lead to respiratory failure and, in up to 40% of cases, death. Spores of the bacteria can reside in soil or in the gastrointestinal tract of animals. C. tetani usually enters a host through a wound to the skin. C. botulinum, C. chauvoei and C. tetani can cause serious diseases in farm animals, resulting in suffering or even death for the animals. Spore-forming bacteria including pathogenic Clostridium species can survive pasteurization and are reportedly not affected by the AD process. Couturier & Galtier report survival of Clostridium following thermophilic digestion. The survival of pathogenic Clostridia during the biogas process results in the need for careful assessment of the risk of application of digested residues onto agriculture. A study by Bagge et al. however failed to show the survival of pathogenic clostridia after AD.

It has been reported that Clostridium can survive the composting process. In a carcass-composting environment, it has been found that anaerobic zones do develop, despite attempts to keep the entire process aerobic. These anaerobic zones allow the growth of Clostridium cells. Eventually, as decomposition occurs, and the temperatures increase, Clostridium growth will slow, and sporulation will occur.

The AH process has been shown to destroy all pathogens listed as index organisms by the State and Territorial Association on Alternative Treatment Technologies (STAATT I and STAATT II), which require a 6-log (99.9999%) reduction in vegetative agents and a 4-log (99.99%) reduction in spore-forming agents. Thus, it could be assumed that Clostridium would be inactivated through AH, and that AH would appear to be the best approach that could be taken to inactivate spore-forming clostridial pathogens.

Brucella Abortus

Bacteria of the genus Brucella are the causative agents for the zoonotic disease brucellosis. B. abortus infects cattle primarily, although it can occasionally also infect sheep. B. abortus infects the placenta and fetus of gestating cows and causes the fetus to abort. When humans however, are infected by this organism they develop a severe fever, but do not abort.

Bacteria of the genus Brucella have been reported to be very heat sensitive. Pasteurization at 63 °C for 30 min and at 72 °C for 15 sec successfully removed Brucella from contaminated raw milk, making it safe for consumption, indicating that pasteurization of animal carcasses could be used to remove the risk of disease. These findings would also suggest that composting could also inactivate Brucella. Similarly, Harada et al. reported that a temperature of 61 °C for 3 min would inactivate Brucella abortis or suis. It is thus likely that a thermophilic AD process would also inactivate this pathogen.

Again, although no literature reports regarding the survival of Brucella after a process of AH could be found based on the report of Kaye et al., it could be expected that AH would be capable of the inactivation of Brucella species.

Bacillus Anthracis

Anthrax is an acute disease caused by Bacillus anthracis. The disease affects both humans and animals, and most forms of the disease are lethal. Infections occur when spores of B. anthracis enter the body via abrasion, inhalation, ingestion, and, on rare occasions, via insect vectors. The endospores are extremely persistent in the environment and can remain viable in soils for up to 300 years.

Reuter et al. conducted an experiment to speculate on the possible fate of B. anthracis within a field scale composting system. Because of regulatory requirements, safety considerations and the field scale design of the experiment, B. anthracis spores were not used, but rather spores of B. licheniformis and B. thuringiensis were chosen. The study showed that a one millionfold inactivation of Bacillus spores was achieved by exposure to compost and elevated temperatures over time. Variation in composting microclimates within the compost pile may explain why some spores remained viable in certain parts of the compost pile. A total inactivation of Bacillus spores was not possible. However, composting may still have merit as a means of bio-containment of infected carcasses and slaughterhouse wastes, reducing and diluting the transfer of infectious spores into the environment.

Bagge et al. investigated the diversity of spore-forming bacteria in cattle manure, slaughterhouse waste and samples from biogas plants. Of particular interest in their study, were Clostridium and Bacillus. Before digested residues are reused for agriculture, biowaste must be adequately treated and proven to be hygienically safe. To reduce the risk of spreading pathogens, pasteurization of biowaste at 70 °C for an hour prior to AD is recommended. Pasteurization however will not inactivate bacterial spores. The results of Bagge et al. showed that Bacillus species seemed to pass the biogas process from biowaste to digested residues relatively unaffected. Due to their survival of pasteurization and the AD process, it is important to assess the risk of spreading digested residues from biogas plants containing Bacillus spores onto agricultural fields, and to take the necessary precautions.

The efficacy of AH for the inactivation of pure cultures of bacteria, including B. subtilis, for the disposal of infectious animal waste was evaluated in a study by Kaye et al. The tissue digester was operated at 110–120 °C and approximately 103 kpa for 18 h. Samples were removed after a cooling phase at 50 °C. The process was found to completely destroy B. subtilis. It can thus be expected that AH would be capable of the inactivation of the more deadly pathogen, B. anthracis.

Mycobacterium Bovis

Mycobacterium bovis is the causative agent of bovine tuberculosis, a chronic disease of cattle and other species. The disease can cause major losses in animal production and is a potentially debilitating zoonosis. M. bovis is capable of jumping the species barrier, and causing tuberculosis in humans, a disease normally caused by M. tuberculosis. Normally, M. bovis is transmitted to humans via infected milk, although it can also spread via aerosol droplets. However, actual infections in humans are rare, because of control measures put in place. Studies by Croshaw et al. showed that M. bovis was destroyed by the pasteurization of milk at 71.7 °C for 15 sec, and at 62.8 °C for 30 min.

Far more work has been conducted investigating the survival of Mycobacterium avium subsp. paratuberculosis after composting than M. bovis. M. avium subsp. paratuberculosis is a member of the same genus, and the causal agent of paratuberculosis, an intestinal infection of domestic and wild ruminants. Grewel et al. showed that after 3 days of composting at 55 °C, M. avium subsp. paratuberculosis were no longer detectable. Also, Gobec et al. showed that M. avium subsp. paratuberculosis was not able to be isolated 24 h after the start of composting. However, a report by Hahesy indicated that farm yard (composted) manure must be exposed to a mean temperature of 60–70 °C during the composting process for 3 weeks to destroy M. bovis bacilli. The authors concluded that as the majority of solid dung heaps do not reach this high a temperature, composted manure cannot necessarily be considered safe. These findings appear to be contradictory of the results of related Mycobacterium studies, and of the findings of pasteurization studies by Croshaw et al.

No studies investigating M. bovis in AD could be found upon intensive literature searches. In a mesophilic AD study by Olsen et al., M. paratuberculosis was detected and cultured up until 28 days of digestion. Under thermophilic conditions, however, M. paratuberculosis was inactivated after 24 h. In a study by Slana et al., viable M. avium subsp. paratuberculosis cells were detected using culture from fermenters up to 2 months. Therefore, the safety of the final products of digestion used for fertilization concerning the presence of Mycobacterium species cannot be certain.

Kaye et al. tested the survival of organisms after AH. All organisms tested, including M. bovis were completely destroyed by the AH process.

Erysipelothrix Rhusiopathiae

Erysipelothrix rhusiopathiae is found all over the world and has been isolated from a variety of vertebrates as well as from soil. E. rhusiopathiae is primarily considered an animal pathogen, causing a disease known as erysipelas in turkeys, pigs, birds, sheep, fish, and reptiles.

Erysipelothrix cultures are reportedly destroyed by exposure to moist heat at 55 °C for 10 min. A study conducted by Morrow investigated composting as an approach to treat swine mortalities.

The authors placed cultures of E. rhusiopathiae in compost piles. The temperatures reached in the composting process were high enough to destroy E. rhusiopathiae, and the authors concluded that properly produced composts did not pose a risk for transmitting disease.

Olsen and Larsen published the results of experiments investigating the reductions of pathogenic and indicator bacteria in animal slurry subjected to mesophilic (35 °C) and thermophilic (53 °C) AD. Under mesophilic AD conditions, a T90 value of 1·8 days was obtained for E. rhusiopathiae, while at 53 °C, the T90 value was reduced to 1–2 h. A study conducted by Han et al. investigated the effects of an anaerobic lagoon fermentation and an autothermal thermophilic aerobic digestion on bacterial pathogens contained in raw swine manure. E. rhusiopathiae was present in the raw manure, however was no longer detectable following a mesophilic anaerobic lagoon fermentation. Manure treated by thermophilic aerobic digestion also did not yield detectable levels of E. rhusiopathiae. These findings suggest that AD processes would be successful in the inactivation of E. rhusiopathiae.

Although no literature reports regarding the survival of E. rhusiopathiae after a process of AH could be found, the report of Kaye et al. would indicate that AH should be capable of the inactivation of E. rhusiopathiae.

BSE

BSE, also known as mad cow disease, is a relatively new disease that primarily affects cattle. BSE can also cause a corresponding disease in humans-Creutzfeldt-Jakob disease. There is still much controversy regarding the causes of BSE and Creutzfeldt-Jakob disease, however the most common belief is that the infectious agents are prions, an abnormal form of a type of protein. However, the prion hypothesis has also been challenged and an autoimmune response theory has been postulated.

Currently, the public considers BSE to be the greatest concern to any bovine-based product. The risk, however, of spreading BSE via composting of catering wastes in the UK has been shown to be 'remote', because there are many controls in place for keeping the disease from entering the food system, and hence the food residuals stream in the UK today. In the slaughterhouse, TSE management aims to prevent infected material from entering the food and feed chains. According to EU legislation, animals suspected of TSE infection are separated and safely disposed of. Prions such as BSE are more resistant to heat than many viruses. In fact, BSE infected material remains infected after cooking, rendering and long periods of incubation in the soil. According to Rohwer, less than a 0.5 log (70%) destruction of scrapie agent was seen after 60 min at 60 °C and 80 °C. It can therefore be assumed that a standard composting process whereby the temperature was maintained at 60 °C for 2 days would not reduce BSE infectivity.

TSEs have also been reported to survive the operational temperatures at which AD are conducted. Studies by Topper et al. showed AD under both mesophilic and thermophilic conditions to be incapable of reducing or eliminating BSE. In a study conducted by Brown et al., infected brain material was heated to 600 °C. Despite brains being totally ashed, when reconstituted with saline to their original weights, ashed brain material was able to transmit disease to 5 of 35 inoculated hamsters.

Apart from incineration (where temperatures of >850 °C are reached); AH is the only effective method known for the destruction of prion material (NABC 2004). A study carried out by the

Institute of Animal Health at the University of Edinburgh, investigated the ability of AH to destroy BSE prions grown in the brains of mice. Infected mice heads were subjected to AH for either 3 or 6 h, and after neutralization of the hydrolysates, aliquots were injected into mice. Evidence of TSE was found in some mouse brains of mice injected with hydrolysate taken from the 3 h digestion, but significantly, no disease was found in the brains of mice injected with hydrolysate from the 6 h digestion. Studies conducted by Murphy et al. also showed scrapie to be inactivated by AH.

Aphto Virus

Foot-and-mouth disease (FMD) is an infectious and sometimes fatal viral disease caused by the Aphto virus that affects domesticated and wild ruminants and pigs. Most infected animals eventually recover but often develop sequelae. The highly contagious nature of the virus and the severity of economic impacts associated with disease have resulted in FMD being labeled as the most important disease limiting the trade of animals and animal products throughout the world.

The virus can survive for weeks or months in refrigerated internal organs, bone marrow, lymph and hemal nodes, glands, and residual blood, however can be inactivated within 3 days in skeletal muscle after slaughter due to reduced pH. The FMD virus can be inactivated using heat, however, the amount of virus inactivation is determined by the temperature, and by access of the heat to the virus infected material. Blackwell et al. reported that heating to an internal temperature of 58.3 °C inactivated FMD in minced beef containing virus-infected lymph nodes, while temperatures of up to 79.4 °C were required for minced meat in nylon tubes. The American Association of Food Hygiene Veterinarians reported that core temperatures of 93 °C were needed to inactivate the FMD virus in heart muscle.

Guan et al. investigated the inactivation and degradation of FMD virus during composting of infected pig carcasses. FMD virus infected pig carcasses were composted in a mixture of chicken manure and wood shavings, and compost temperatures reached 50 °C and 70 °C by days 10 and 19, respectively. Results indicated that the virus was inactivated in specimens in compost by day 10 and the viral RNA was degraded in skin and internal organ tissues by day 21.

Inactivation experiments conducted by Turner et al. in contrast, showed the FMD virus to be remarkably thermally stable, surviving with little or no loss of titre at 55 °C and 60 °C. Even at 65 °C, the virus did not drop to below detectable levels within 5 min in a slurry. From these studies, the authors recommended treatment of slurry and water-containing FMD virus in full-scale mobile treatment plant operating at a minimum of 70 °C.

As with other viruses, the bone marrow of infected animal carcasses can contain high loadings of FMD virus. Therefore, to effectively inactivate FMD virus from animal waste material, a composting process where the temperature was maintained at greater than 65 °C for at least 2 days would need to be used. According to Gale et al., a composting temperature of 60 °C for 2 days would give at least a 6-log reduction of all exotic pig viruses, including FMD, even in the bone marrow of whole pig carcasses discarded for composting.

No information could be found regarding the survival of FMD virus after either mesophilic or thermophilic AD. Considering the heat stability of the virus, it could be predicted that AD would not successfully inactivate FMD virus. However, the combination of a pasteurization step and an AD process could be expected to be successful in the inactivation of the virus.

No literature reports regarding the survival of FMD virus after a process of AH could be found. As the report of Kaye et al. states that the process was found to completely destroy all representative classes of potentially infectious agents, it can be expected that AH would be capable of the inactivation of this virus.

Lyssa Virus

Rabies is a disease caused by the lyssa virus, and is an acute encephalitis in warm-blooded animals. In humans, rabies is almost always fatal if post-exposure prophylaxis is not administered prior to the onset of severe symptoms. The economic impact of the disease is significant, with the virus causing death of livestock in some countries.

The rabies virus is able to be inactivated by heat and thus a pasteurization step at 70 °C should be considered successful in rabies inactivation. No information was able to be found on composting inactivation of the virus, but based on the findings of Kopcha et al., it would be likely that composting would also successfully inactivate the virus. Studies by Kissling and Reese investigated virus stability at different temperatures. Rabies tissue culture-adapted virus in serum-free medium was reported to be sensitive to heat inactivation. At 56 °C, almost all virus was inactivated within 30 min, while at 37 °C, more than 99 % of the virus was inactivated within 24 h. At 37 °C, no virus was viable after 32–48 h.

No information could be found upon extensive literature searching on the survival of the rabies virus after an AD process. Considering the heat sensitivity of the virus, it is likely that it would be inactivated after a thermophilic AD, but it is not certain whether a mesophilic AD process would have the same effect. When combined with a pasteurization step however, one could hypothesize that the virus would be inactivated.

Again, no literature reports regarding the survival of the rabies virus after a process of AH could be found. Based on the findings of Kaye et al., it would however be expected that inactivation of the virus would result from AH.

African Swine Fever Virus

African swine fever (ASF) is a viral disease (genus Asfivirus) which causes a devastating haemorrhagic fever of pigs with mortality rates approaching 100%. Virus is excreted in the faeces of infected pigs, and as faeces mostly are transferred to a slurry store prior to the infection being identified, viral contamination of large quantities of pig slurry can occur inadvertently.

ASF virus has been reported to be quite resistant to inactivation, and can survive many freeze-thaw cycles. It is also resistant to changes in pH, and certain strains are resistant to complete inactivation at pH values between 4 and 13. The virus is however very sensitive to drying, and can be inactivated by lipid solvents because of its lipid envelope.

ASF virus is however heat sensitive. In studies conducted by Plowright and Parker, ASF virus was shown to be inactivated (5 log inactivation) by 90 min at 56 °C. However, work done by Turner et al. indicated the virus to be inactivated at 50 °C in 30 min, at 56 °C within 90 sec and within seconds at 60 °C.

The findings of Turner et al. described above, as well as projected costs of treatment for virus inactivation resulted in the authors deciding that heat was the most effective means of ASF virus inactivation. This is because heat treatment was found to be very rapid and consistent at a temperature of 56 °C, while alkali inactivation was slower. Also, heat can be recovered, thus reducing costs. These authors designed a pilot scale treatment plant to treat pig slurry at a minimum temperature of 65 °C, and maintained this temperature in at least 99.99% of the material for a minimum of 5 min. This was done so to provide a reasonable safety margin for virus inactivation.

The bone marrow of infected pig carcasses contains high loadings of ASF virus. When considering the composting of an animal carcass, the times and temperature of the process must be carefully considered. It has been estimated that a sphere of diameter 40 cm (which would be similar to a large leg or carcass of pork containing the bone) would require 40 h to reach a temperature of 56 °C at the centre when the surrounding composting temperature is 60 °C. Therefore, a composting process with a temperature of 60 °C maintained for 2 days would give at least a 6-log reduction of all exotic pig viruses, including ASF virus, even in the bone marrow of whole pig carcasses. As studies have shown ASF virus to be inactivated within seconds at 60 °C, it would appear that a correctly performed composting process should be successful in inactivating ASF virus.

No information could be found regarding AD and the survival rates of ASF virus subjected to the process. While no predictions can be made as to the survival of the virus in a mesophilic process, it could be predicted that a thermophilic process may successfully inactivate the virus.

A 1% solution of both NaOH or $Ca(OH)_2$ was found to be able to result in a 4 fold inactivation of the virus at 4 °C within 150 sec. More dilute solutions (0.5%) of both chemicals were however not found to be able to inactivate virus at 22 °C in 5 min. It could be expected that a AH process, which uses alkaline solution as well as heat to inactivate microbial organisms would be successful in the inactivation of ASF virus.

Phlebo Virus

Rift Valley fever (RVF) is a viral zoonosis resulting in fever. The disease is caused by the RVF virus, a member of the Phlebovirus genus. It is spread by the bite of infected mosquitoes, or by exposure to tissues or blood of infected animals. Humans infected with the virus have self-limited febrile illness, but retinal degeneration (5–10%), hemorrhagic fever (<1%), or encephalitis (<1%) can also develop. The RVF virus is susceptible to low pH (≤6.2), lipid solvents and detergents, as well as solutions of sodium or calcium hypochlorite with residual chlorine content greater than 5000 ppm. In a neutral or alkaline pH, mixed with serum or other proteins, the virus can survive for as long as four months at 4 °C and eight years below 0 °C. The virus is also quickly destroyed by pH changes in decomposing carcasses. Thus, although no literature can be found on AH of animals infected with RVF virus, it would appear that AH would not be a recommended method of treatment of animal carcasses in order to inactivate the virus.

Unfortunately, no information could be found on composting or AD with respect to the inactivation of RVF virus, although Flory et al. report that composting approaches can be used to treat RVF diseased animals, thus preventing disease outbreaks.

Cysticercus Bovis

Bovine cysticercosis is a zoonotic disease found worldwide affecting the muscles of cattle. The disease is most prevalent in countries where poor sanitation practices on cattle farms are common and where cultural habits include eating undercooked meat. It does not represent a serious human health risk. Bovine cysticercosis is caused by Cysticercus bovis, the larval stage of Taenia saginata, a species of human tapeworm. Humans can acquire intestinal tapeworm infections by the consumption of beef containing the parasite. Disease in cattle is initiated by the ingestion of materials contaminated with tapeworm eggs originating from human feces larvae of a human tapeworm. Cattle become infected with bovine cysticercosis by ingesting materials contaminated with tapeworm eggs originating from human faeces.

A study was conducted by Hughes et al. to investigate the survival of T. saginata ova after AD at 35 °C for 10 days followed by lagooning for 15 days. Results indicated that viability and infectivity were completely destroyed. The authors also showed that a 3 h treatment at 55 °C in crude sludge resulted in a 99 % reduction of ova infectivity. It could thus be assumed that a 70 °C pasteurization step prior to AD treatment, as is the regulation in Europe, would successfully inactivate bovine cysticercosis.

Another study by Morris et al. showed 100% efficacy of sanitation of T. saginata at 35 °C. A thorough review of the literature has failed to find studies whereby the inactivation of T. saginata by composting was investigated. However, the US Environmental Protection Agency (Part 503) sets standards for the destruction and concentration of pathogens in composted biosolids. These standards are commonly used as marketplace specifications for compost sold for public use. By maintaining these standards, various harmful pathogens should be destroyed. According to the standards, T. saginata does not survive for more than a few minutes at 55 °C. Thus, a standard composting procedure would successfully inactivate this organism.

No information could be obtained regarding the inactivation efficiency of AH on T. saginata. However, Kaye et al. showed Giardia cysts to be completely destroyed by AH, an indication that slaughterhouse waste material infected by bovine cysticercosis might also successfully be treated by AH.

HAZARDOUS WASTE MANAGEMENT

Hazardous Waste Management is a process that involves the collection, recycling, treatment, transportation, disposal, and monitoring of wastes disposal sites. Currently, in most developing and transition countries, hazardous wastes often are disposed off indiscriminately posing health and environmental risk. Increasingly, governments and international agencies are attempting to control the growing problem of hazardous substances in the environment which proves difficult because there are so many sources through which they are released.

So many toxic and hazardous substances emanate from these sources either purposefully or accidentally contaminating, the land, air, and water. The potential health effects of these substances range from minor, short term discomforts, such as headaches and nausea to serious health problems, such as cancers and birth defects (that may not manifest themselves over years), to major accidents that cause immediate injury or death.

According to Roger et al., proper control of hazardous wastes does cost money, but experience in a number of developed countries suggests that cleaning up the "sins of the past" is much more expensive in the long term. For instance, in the United States, cleanup of improperly managed wastes has been estimated to cost 10-100 times as much as proper early management. It is therefore important that all developing countries institute controls over hazardous wastes to avoid excessive costs in the future.

In view of this, management of hazardous wastes including their disposal in environment friendly and economically viable way is very important and therefore suggestions are made considering the waste types and states (solid or liquid).

Sources of Hazardous Wastes

Most hazardous wastes are produced from a wide range of sector mainly in the manufacturing of products for consumption or further industrial application. These include and not limited to: Agricultural land and agro industry, domestic/household, mines and mineral processing sites, health care facilities, commercial sites, institutional, industrial, solid waste disposal sites, contaminated sites, building materials and activities that involve radioactive elements.

Table shows the widespread distribution of hazardous wastes. Many sectors of the economy are victims of hazardous wastes production whether directly and indirectly. Some industries lack the capacity and capability to properly handle these wastes; they resort to indiscrimate disposal as the only way to avoid costs and procedures associated with proper management.

Sector	Source	Hazardous Wastes
Agricultural land and agro-industry	Farms	Pesticides, fertilizers, and hazardous veterinary product wastes
Health care centers	Hospitals, clinics and medical laboratory	Pathological wastes, infectious needles, contaminated blood.
Commercial wastes	Gasoline stations, dry cleaners and automobile repair and servicing shops (workshops), airports, municipal parks, Electrical transformers	Waste oils, hydraulic fluids, halogenated solvents, polychlorinated Biphenyls.
Industrial Hazardous Wastes	Petroleum fuel	Refining products (fuels and tar), impurities like phenol and cyanides in the waste stream, and sludge flushed from the storage tanks.
Small scale industry	Metal treating (electro- plating, etching, anodizing, galvanizing), Photofinishing Textile processing, Printing, Leather tanning.	Acids, heavy metals Solvents, acids, silver Cadmium, mineral acids Solvents, inks and dyes Solvents, chromium
Large Scale industry	Bauxite processing, Oil refining, Petrochemical manufacture, Chemical/pharmaceutical manufacture, Chlorine production	Red muds Spent catalysts Oily wastes Tarry residues, solvent Mercury.
Building and Roofing Industries	Building and roofing materials	Materials made of asbestos and copper
Mines and Mineral processing, Electricity generating companies	Uranium mining and processing	Spent reactor rods, cloths, rags and wood

Some of these substances are classified as probable human carcinogens and are associated with increased tumor rates, organ damage and even death in man.

Effects of Hazardous Substances on the Environment

Today, because of increased urbanization and industrialization, some communities are threatened with pollution caused by improper disposal of hazardous wastes (solid/liquid), and these wastes can adversely affect the physical, mental and social health of the inhabitants.

Effects on Land

Pollution resulting generally as a result of man's activities affects both crops, animal and human health, and also the quality of the whole environment. Thus it is any process carried out by man which leads to an adverse change in terrestrial environment. The dumping of refuse could be termed the defacement of land and subsequent rendering of the soil nonproductive and polluted. Hazardous wastes dumped on land sites, when it rains, the leachates from such refuse dumps contain various contaminants, the most harmful being heavy metals, dissolved organic compounds, and nitrogen.

Deteriorating soil quality and decrease in vegetation abundance are also grave consequences of open waste dumping which have resulted in growing public concern. Figures below show indiscriminate ancient disposal methods called "open dumps". Open dumps in addition to other effects mentioned earlier lead to deteriorating soil quality and decrease in vegetation abundance.

Environmental contamination by components of refuse such as heavy metals like cadmium, lead, zinc and copper has been of great concern in the last decades because of their health hazards to man and other organisms; when accumulated within a biological system. The response of plants to pollutants is especially important since many of these plants are more sensitive than the organisms.

Open dumps-Mixed wastes.

Open dumps-empty used oil drums.

This is why plants usually serve as the most sensitive indicators as prolonged exposure to them may destroy a large number of plants communities. Toxicity can result from a high concentration of an element nutrient intake and can also give rise to abnormal growth, disease or even death.

Also refuse dumps serve as breeding places for diseases-causing organisms such as rats and flies which are the major disease vectors associated with garbage. Rats, with the assistance of flies are known to be involved in the spread of bubonic plague and urine typhoid fever to humans. Rats may contaminate food with their urine, transmitting leptospirosis, lasa fever and infectious hepatitis. They can transmit salmonella and other intestinal organisms.

Also, hazardous wastes spoil the natural beauty of the land and may interfere with recreational activities. Hazardous solid wastes can also produce unpleasant smell. The free flow of traffic along our streets could also be hindered by the indiscriminate dumping of solid hazardous wastes and can occupy land space that could have been used for other purposes as seen in figures. Solid waste has the potential to cause sickness and even death. Therefore proper waste management in any society is a matter of urgent attention.

Inherent changes in the application of industrial sludge to land due to contaminations with heavy metals and possible presence of pathogens in raw sludges cause its prohibition in agricultural lands. Chronic cadmium poisoning causes damage to kidney and heart. Prolong exposure also results in loss of calcium from the bones, which then become brittle and easily break. In Japan for example, a whole village was striken by cadmium poisoning-known as "Itai-Itai" disease after effluents contaminated rice supplies. Other heavy metals and hazardous substances of interest are Copper, Chromium, Zinc, Lead, Arsenic, and Mercury and Selenium, etc and their effects on man are shown on table below.

Heavy Metals/Substances	Source	Toxicity
Cadmium (Cd)	Textile Processing	Renal and heart diseases, weakens bonesand tumors.
Lead (Pb)	Mines and mineral processing companies	Neurological disorders, convulsions, kidney and brain damage
Mercury (Hg)	Chlorine Production	Neurological disorders, irritability and liver damage
Arsenic (As)	Metal treating (electro-plating, etching, anodizing, galvanizing)	Cancer, skin diseases and poisoning
Zinc (Zn)	Building and roofing materials	Enzymes and skin disorders
Copper (Cu)	Building materials made of copper	Gastrointestinal irritant, liver damage
Selenium (Se)	Photographic devices, gun cleaning solutions, plastics, paints, anti- dandruff shampoos, vitamin and mineral supplements, fungicides and some types of glass.	Hair loss, nervous disorders.
Vinyl chloride	Plastics manufacturing. Air or water at contaminated sites	Acute effects: dizziness, headache, unconsciousness, death Chronic effects: liver, lung, and circulatory damage
Polychlorinated biphenyls (PCB)	Eating contaminated fish, Industrial exposure	Probable carcinogens; acne and skin lesions
Benzene	Industrial exposures, Glues, cleaning products, gasoline	Acute effects: drowsiness, headache, death at high levels. Chronic effects: damage to blood-forming tissues and immune system; also carcinogenic

Benzoapyrene	Product of combustion of gasoline or other fuels. In smoke and soot	Probable carcinogen; probable birth defects
Polycyclic aromatic hydrocarbons	Exposure to smoke from a variety of sources	Probable carcinogen; possible birth defects
Benzobfluoranthene	Product of combustion of gasoline and other fuels inhaled in smoke	Probable carcinogen
Chloroform	Contaminated air and water. Many kinds of industrial settings	Affects central nervous system, liver, and kidneys; probable carcinogen
DDT	From food with low levels of contamination. Still used as pesticide in parts of world	Probable carcinogen, possible long term effect on liver; possible reproductive problems.
Trichloroethylene	Used as a degreaser, evaporates into air	Dizziness, numbness, unconsciousness, death
Dibenza, hanthracene	Product of combustion in smoke	Probable carcinogen

Effects on Water

It is a generally accepted fact that the developed countries suffer from problems of chemical discharge into water sources mainly groundwater, while developing countries face problems of agricultural run-off in water sources. On top of these concerns, contamination of water is a big concern after a natural disaster (e.g., tsunamis, earthquakes, hurricanes, floods, and volcanoes) and can influence water quality on a grand scale.

Secure landfills may not be totally secure even if conscientiously designed. Carefully compacted clay may be very low in permeability but is probably never completely impermeable, especially over long time intervals. Chemical and biological reactions in the wastes and leachate can rupture or decompose plastic, and the stress caused by the weight of wastes and cover can fracture a clay liner. Leakage from "secure" toxic waste dumps, in which hundreds or thousands of barrels of concentrated toxic liquid chemicals are stored, has far more potential for harm. Leakage of at least certain volatile organic compounds deemed high priority toxic pollutants like benzene, toluene, vinyl chloride and chloroform have been detected by monitoring wells, e.g "Mount Trashmore" Landfill in Evanston, Illinois. These pollutants have been confirmed to cause adverse health problems to man. Prevalent high rainfall and shallow water table accentuate problems of pollution from landfills.

Our water supply is threatened with the presence of asbestos particles, heavy metals like Lead, Cadmium, Arsenic, nitrates and sodium, a variety of chemicals that are known to be carcinogens. Almost all the millions of tons of chemicals improperly disposed off on land will sooner or later find their way to the underground water aquifers if they are fully biograded. These chemicals include toxic synthetic organic chemicals developed with the explosive growth of chemicals and petrochemicals since World War 11 such as Trichloroethane, tetrachloroethane, venylchloride, 1,1,2 - trichloroethane, carbon tetrachloride, benzene, xylene, toluene, and methylene chloride.

Some of these substances are essential to the body in trace amounts, mainly for synthesizing enzymes, but it can be toxic in large amounts. According to the Doctor's complete guide to Vitamins and minerals, "the tolerable upper intake level of selenium is set at 400 mg daily, with 800 mg daily potentially causing toxicity and a dose of 5 mg considered lethal for most people. Selenium poisoning has no known antidote, so recognition of the signs and symptoms is essential for survival.

Most of these chemicals are petroleum based and are found in gasoline (petrol) septic tank cleaners, deodorants, plastics, spot removers, household cleaners, disinfectants, paints and varnish removers, dry cleaning fluids and degreasing agent etc. These polluting chemicals find their ways to the ground water in many ways-principally by seeping into aquifers from landfills and waste dumps where chemicals are routinely dumped. They also come from septic tanks of sub-urban, urban and rural homes and from ruptured homes, underground petrol and fuel oil storage tanks. There is no easy solution to the polluted aquifers. Restoration is almost impossible; once contaminated, the ground water remains so. Once the chemicals reach the deep recesses of aquifers, they congeal together and their pollutants grow in size and height yearly depending on the shape and composition of the geological structures in the aquifer.

The solution to this problem is that of drilling into deeper less contaminated aquifer. There is no guarantee that the toxic substances in aquifer near the surface will not reach the lower ones. Aeration of the water is another method. Hence the contaminated water is drawn up from the aquifers, passed through a cooling tower and then sprayed by nozzles into the air. In this way, many of the volatile toxic chemicals evaporate before the water hits the ground again and seeps into the aquifer.

Cancers, arthritis, skin irritation, heart disease, damage to the central nervous system, skin rashes, kidney problems and bronchitis etc., are the diseases associated with water pollution by chemicals. The amount of contaminated water that can cause sickness to its consumer depends on concentration and composition of the contaminant chemicals.

Handling of Hazardous Wastes

Persons handling hazardous wastes are advised to wear respiratory protection, hand and skin protection, eye protection and protective clothing to reduce the effect of contact. Hazardous wastes can cause dermatitis to the skin, some cause asthma on long exposure, others cause the eyes to smart and run and also tightening of the chest.

Transport of Hazardous Waste

Hazardous waste generated at a particular site often requires transport to an approved treatment, storage, or disposal facility (TSDF). Because of potential threats to public safety and the environment, transport is given special attention by governmental agencies. In addition to the occasional accidental spill, hazardous waste has, in the past, been intentionally spilled or abandoned at random locations in a practice known as "midnight dumping." This practice has been greatly curtailed by the enactment of laws that require proper labeling, transport, and tracking of all hazardous wastes.

When hazardous wastes are transported off-site for disposal, the waste generator prepares a shipping document called a manifest. This form must accompany the waste to its final destination and used to track the waste"s movements from "cradle to grave".

In the event of a leak or accidental spill of hazardous waste during its transport, the transporter must take immediate and appropriate actions, including notifying local authorities of the discharge. An area may have to be diked to contain the wastes, and efforts must be undertaken to remove the wastes and reduce environmental or public health hazards.

Waste Minimization and Recycling

Waste minimization is a very important hazardous waste management strategy in developing countries. The concept of waste minimization can best be illustrated by means of figure which subdivides the strategy into three headings:

Hazardous Waste Generation Process

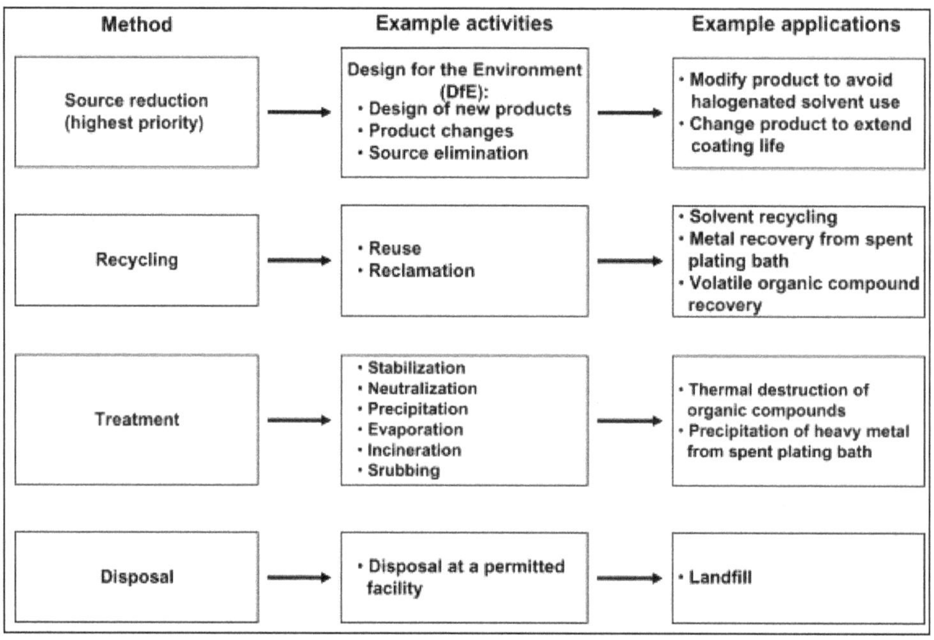

Hazardous Waste Control through Reduction.

- Volume reduction –Preferable accomplished without an increase of toxicity.
- Toxicity reduction –Preferable accomplished by means other than dilution.

Recycling and recovery: Viewed generically "recycling" encompasses both reuse and reclamation activities. Many hazardous wastes can be recycled into new products.

Examples might include lead-acid batteries or electronic circuit boards, where there are heavy metals, after burning, these types of ashes go through the proper treatment, they could bind to other pollutants and convert them into easier-to- dispose solids, or they could be used as pavement filling. Such treatments reduce the level of threat of harmful chemicals, like fly and bottom ash while also recycling the safe product.

Hazardous waste recycling is divided into two namely; materials recovery as well as energy recovery. A recycler's decision as to how to treat a waste is principally determined by the character of specific waste streams or waste mixtures. Where treatment should take place (either onsite or offsite however, is a function of a generator's management practices which include:

- Proximity to offsite recycling facilities.
- Economic costs related to the transportation of wastes.
- The volume of wastes available for processing, and
- Costs related to storage of waste onsite to offsite.

Recycling is characterized by three major practices:

- Direct use or re-use of a waste in a process.
- Recovery of a secondary material for a separate end use such as the recovery of a metal from a sludge.
- Removal of impurities from a waste to obtain a relatively pure re-usable substance.

Although recycling of selected wastes is practiced to a considerable degree by certain industries, only a small percentage of the hazardous waste is being recycled. A greater percentage of wastes are being recycled onsite. Offsite recycling however has become increasingly common with the advent of commercial recyclers and direct transfer of waste from generators to others who can re-use the wastes. Most flammable materials can be recycled, used as industrial fuel. Some materials with hazardous constituents can be recycled, lead acid batteries are one example.

Recycled wastes are used as feedstocks in production processes or as substitutes for commercial chemical products. Examples include:

- The re-use of solvents for equipment cleaning;
- The recycling of collected pesticide dusts at pesticides formulators; and
- The re-use of ferric chloride wastes from titanium dioxide manufacturing as a wastewater conditioner in water treatment.

In the 1981 U.S. survey, data indicate that recycling for materials recovery and re-use appears to be more popular than fuel use or energy recovery. Reasons are, first, some wastes that could be recycled for energy recovery can also be reclaimed and re-used over and over.

Energy recovery in contrast destroys the inputs. Only when the waste is too "dirty" (contaminated from repeated re-use) do operators consider energy recovery a desirable option. The 1981 data may not, however, provide a completely accurate picture of current practices because of recent developments in energy recovery technology. Many technologies were not available in 1981, and others are only beginning to be commercially available today.

Solvents tend to be used for energy recovery because they can possess high energy values. Increasing quantities of high calorific wastes are being used by cement plants and lime kilns.

Waste Treatment and Disposal

Treatment

While reduction and recycling are desirable options, they are not regarded as the final remedy to the problem of hazardous-waste disposal. There will always be a need for treatment, and for storage before the final disposal of some amount of hazardous waste. Hazardous waste can be treated by chemical, thermal, biological, and physical methods.

Hazardous waste is the most difficult waste to manage, since in the treatment process, heavy metal and dioxin among others are obtained. The outcome elements are dangerous not only for the environment but also for public health.

Disposal

Disposal of hazardous waste is the final stage of a hazardous waste management system. Almost all possible disposal methods necessitate proper pre-treatment, in order to have secure disposal facilities. Synopsis of selected principles of different waste disposal methods are emphasized in this work namely: Secure landfill, Deep well and Bedrock disposal.

Secure Landfills

Historically, some hazardous wastes were disposed of in regular landfills. This resulted in unfavorable amounts of hazardous materials seeping into the ground. These chemicals eventually enter natural hydrologic systems. Many landfills now require countermeasures against groundwater contamination, an example being installing a barrier along the foundation of the landfill to contain the hazardous substances that may remain in the disposed waste. Currently, hazardous wastes must often be stabilized and solidified in order to enter a landfill and many hazardous wastes undergo different treatments in order to stabilize and dispose of them.

A landfill is a disposal facility where hazardous wastes are placed into and stored in the soil. Landfills for hazardous wastes are frequently considered a technology of last resort to be used after every effort has been made to reduce or eliminate the hazards posed by the waste. The intent is to bury or alter the wastes so that they are not environmental or public health hazards.

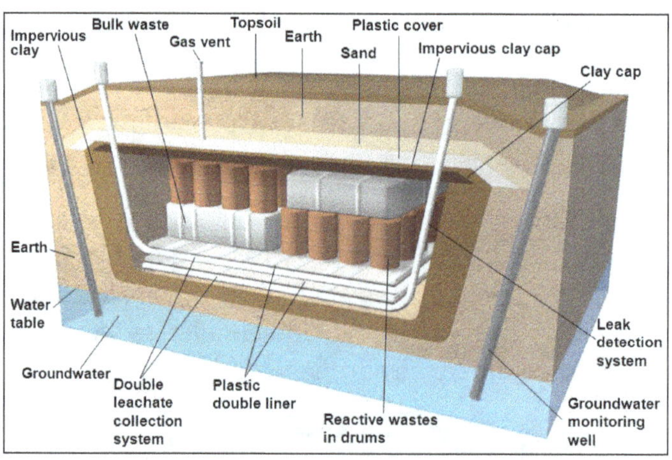

A secure landfill design for hazardous waste disposal.

For some years, waste disposal specialists have believed that, in principle, it is possible to design a secure landfill site for hazardous liquid wastes. An example of a recommended design is shown in figure. The wastes are put in sealed drums before disposal. A secure hazardous-waste landfill must have two impermeable liners and leachate collection systems.

The double leachate collection system consists of a network of perforated pipes placed above each liner. The upper system prevents the accumulation of leachate trapped in the fill, and the lower serves as a backup. Collected leachate is pumped to a treatment plant. In order to reduce the amount of leachate in the fill and minimize the potential for environmental damage, an impermeable cap or cover is placed over a finished landfill.

In view of the above precaution to prevent leachates from accumulating in landfills, there isn't any complete guarantee that leachate would not accumulate no matter how secure the landfill is engineered.

The main components in the leachate from landfill sites may be conveniently grouped into four classes, as follows:

- Major elements and ions such as calcium, magnesium, iron, sodium, ammonia, carbonate, sulphate and chloride;
- Trace metals such as mangenese, chromium, nickel, lead, and calcium;
- A wide variety of organic compounds which are usually measured as Total Organic Carbon (TOC) or Chemical Oxygen Demand (COD); individual organic species such as phenol can also be of concern;
- Microbiological components.

All hazardous waste and most industrial waste will give rise to leachate.

Heavy metals have been of greater concern compared to other leachate components; this is as a result of its persistent and bio-accumulative nature in the food chain. Fossil fuel combustion, metallurgical industrial activities and industrial wastes generation are the main sources of heavy metals' wastes generation, and also poor management by generators over years have accentuated their unwarranted release to the environment.

Deep-well Disposal

Another alternative disposal of liquid industrial waste is injection into deep wells. Deep well injection is a liquid waste disposal technology.

This alternative uses injection wells to place treated or untreated liquid waste into geologic formations that have no potential to allow migration of contaminants into potential potable water aquifers. This method has been practiced since World War 2 and involves pumping liquid waste through a steel casing into a porous layer of limestone or sandstone. High pressures are applied to force the liquid into the pores and fissures of the rock, where it is to be permanently stored. The rock unit selected to receive the waste must be relatively porous and permeable (commonly, sandstone or fractured limestone), and it must be isolated by lowpermeability layers (for example, shale) above and below. The subsurface geology must be known in sufficient detail that there is reasonable confidence that the disposal stratum remains isolated for some distance from the well site in all directions. Information about that geology may be derived from many sources: direct drilling to obtain core samples that provide a vertical section of the rock units present; geophysical studies that provide data on depths to and thicknesses of different rock layers, and on distribution of ground water; geologic mapping on the basis of cores, surface outcrops, and geophysical data to interpolate between points sampled directly.

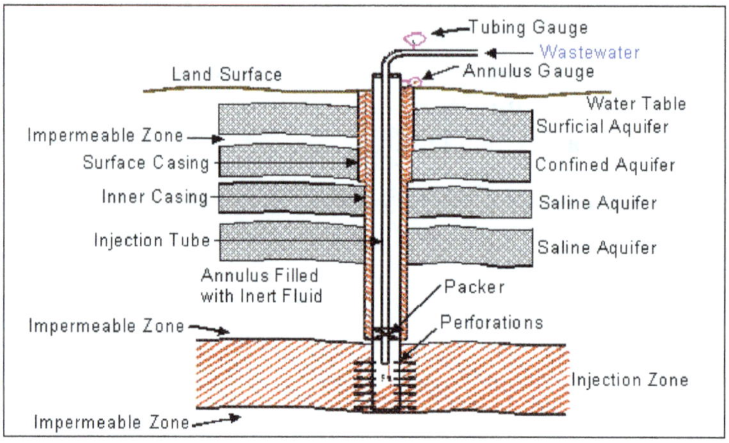

Deep well disposal method.

These disposal wells are hundreds to thousands of meters deep, far removed from the surface, and below the regional water table. The pore water in the disposal stratum should be brackish or saline water not suitable for a water supply. Where the well intersects any shallower aquifer that are or might be used for water supply, it must be snugly lined (cased) to prevent leakage of the wastes into those aquifers. Local well water is monitored to detect any accidental leaks promptly.

Movement of deep ground water is generally slow, and the assumption is that, by the time the toxic chemicals have migrated far enough laterally to reach a usable aquifer or another body of water, they will have become sufficiently diluted not to pose a threat. This presumes knowledge of the toxicity of the chemicals in low concentrations. When the wastes are more or less dense than the ground water, they displace and not miscible with it, folds or other geologic structure may help to contain them and slow their spread, as oil traps contain petroleum. The behaviour of chemicals that dissolve in pore water is much less well understood. They can diffuse through the water more rapidly than the water itself moves, so that even if deep ground water transport is slow, contaminant migration may not be.

Deep-well injection is relatively inexpensive and requires little or no pretreatment of the waste, but it poses a danger of leaking hazardous waste and eventually polluting subsurface water supplies.

Bedrock Disposal

A variety of bedrock types are being investigated as host rocks for solid hazardous waste. The general design of a bedrock disposal site or repository for such wastes is shown in figure. It involves the multiple barrier (or multi barrier) concept: surrounding solid waste with several different types of materials to create multiple obstructions to waste leakage or invasion by ground water. A major variable is the nature of the host rock as well as some potential drawbacks. The waste forms envisaged for disposal are vitrified high-level radioactive wastes sealed into stainless steel canisters, or spent fuel rods encapsulated in corrosion resistant metals such as copper or stainless steel. The most widely accepted plans are for these to be buried in stable rock structures deep underground. Many geological formations such as granite, volcanic tuff, salt, thick basalts such as the Columbia River plateau basalt or shale will be suitable.

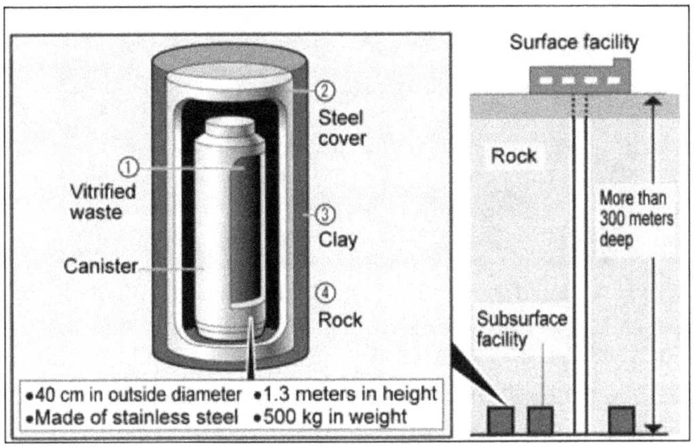

Bedrock disposal of solid hazardous wastes.

The fundamental goal of hazardous waste management program is to change the behavior of those who generate hazardous waste so that they routinely store, transport, treat, and dispose of them in an environmentally safe manner. The focus comes in an effort to address more immediate threats to public health, such as safe drinking water.

Hazardous waste management must move beyond burying and burning. Industries need to be encouraged to generate less hazardous waste in their manufacturing processes. Although toxic wastes cannot be entirely eliminated, technologies are available for minimizing, recycling, and treating wastes. It is possible to enjoy the benefits of modern technology while avoiding the consequences of a poisoned environment. The final outcome rests with governmental and agency policy makers, as well as with an educated public.

HIGH-LEVEL RADIOACTIVE WASTE MANAGEMENT

High-level radioactive waste management concerns how radioactive materials created during production of nuclear power and nuclear weapons are dealt with. Radioactive waste contains a mixture of short-lived and long-lived nuclides, as well as non-radioactive nuclides. There was reported some 47,000 tonnes of high-level nuclear waste stored in the USA in 2002.

The most troublesome transuranic elements in spent fuel are neptunium-237 (half-life two million years) and plutonium-239 (half-life 24,000 years). Consequently, high-level radioactive waste requires sophisticated treatment and management to successfully isolate it from the biosphere. This usually necessitates treatment, followed by a long-term management strategy involving permanent storage, disposal or transformation of the waste into a non-toxic form. Radioactive decay follows the half-life rule, which means that the rate of decay is inversely proportional to the duration of decay. In other words, the radiation from a long-lived isotope like iodine-129 will be much less intense than that of short-lived isotope like iodine-131.

Spent nuclear fuel stored underwater and uncapped at the Hanford site in Washington, USA.

Governments around the world are considering a range of waste management and disposal options, usually involving deep-geologic placement, although there has been limited progress toward implementing long-term waste management solutions. This is partly because the timeframes in question when dealing with radioactive waste range from 10,000 to millions of years, according to studies based on the effect of estimated radiation doses.

Thus, engineer and physicist Hannes Alfvén identified two fundamental prerequisites for effective management of high-level radioactive waste: (1) stable geological formations, and (2) stable human institutions over hundreds of thousands of years. As Alfvén suggests, no known human civilization has ever endured for so long, and no geologic formation of adequate size for a permanent radioactive waste repository has yet been discovered that has been stable for so long a period. Nevertheless, avoiding confronting the risks associated with managing radioactive wastes may create countervailing risks of greater magnitude. Radioactive waste management is an example of policy analysis that requires special attention to ethical concerns, examined in the light of uncertainty and futurity: consideration of 'the impacts of practices and technologies on future generations'.

There is a debate over what should constitute an acceptable scientific and engineering foundation for proceeding with radioactive waste disposal strategies. There are those who have argued, on the basis of complex geochemical simulation models, that relinquishing control over radioactive materials to geohydrologic processes at repository closure is an acceptable risk. They maintain that so-called "natural analogues" inhibit subterranean movement of radionuclides, making disposal of radioactive wastes in stable geologic formations unnecessary. However, existing models of these processes are empirically underdetermined: due to the subterranean nature of such processes in solid geologic formations, the accuracy of computer simulation models has not been verified by empirical observation, certainly not over periods of time equivalent to the lethal half-lives of high-level radioactive waste. On the other hand, some insist deep geologic repositories in stable geologic formations are necessary. National management plans of various countries display a variety of approaches to resolving this debate.

Researchers suggest that forecasts of health detriment for such long periods should be examined critically. Practical studies only consider up to 100 years as far as effective planning and cost evaluations are concerned. Long term behaviour of radioactive wastes remains a subject for ongoing research. Management strategies and implementation plans of several representative national governments are described below.

Geologic Disposal

The International Panel on Fissile Materials has said:

> It is widely accepted that spent nuclear fuel and high-level reprocessing and plutonium wastes require well-designed storage for periods ranging from tens of thousands to a million years, to minimize releases of the contained radioactivity into the environment. Safeguards are also required to ensure that neither plutonium nor highly enriched uranium is diverted to weapon use. There is general agreement that placing spent nuclear fuel in repositories hundreds of meters below the surface would be safer than indefinite storage of spent fuel on the surface.

The process of selecting appropriate permanent repositories for high level waste and spent fuel is now under way in several countries with the first expected to be commissioned some time after 2017. The basic concept is to locate a large, stable geologic formation and use mining technology to excavate a tunnel, or large-bore tunnel boring machines (similar to those used to drill the Channel Tunnel from England to France) to drill a shaft 500–1,000 meters below the surface where rooms or vaults can be excavated for disposal of high-level radioactive waste. The goal is to permanently isolate nuclear waste from the human environment. However, many people remain uncomfortable with the immediate stewardship cessation of this disposal system, suggesting perpetual management and monitoring would be more prudent.

Because some radioactive species have half-lives longer than one million years, even very low container leakage and radionuclide migration rates must be taken into account. Moreover, it may require more than one half-life until some nuclear materials lose enough radioactivity to no longer be lethal to living organisms. A 1983 review of the Swedish radioactive waste disposal program by the National Academy of Sciences found that country's estimate of several hundred thousand years—perhaps up to one million years—being necessary for waste isolation "fully justified."

The proposed land-based subductive waste disposal method would dispose of nuclear waste in a subduction zone accessed from land, and therefore is not prohibited by international agreement. This method has been described as a viable means of disposing of radioactive waste, and as a state-of-the-art nuclear waste disposal technology.

In nature, sixteen repositories were discovered at the Oklo mine in Gabon where natural nuclear fission reactions took place 1.7 billion years ago. The fission products in these natural formations were found to have moved less than 10 ft (3 m) over this period, though the lack of movement may be due more to retention in the uraninite structure than to insolubility and sorption from moving ground water; uraninite crystals are better preserved here than those in spent fuel rods because of a less complete nuclear reaction, so that reaction products would be less accessible to groundwater attack.

Materials for Geological Disposal

In order to store the high level radioactive waste in long-term geological depositories, specific waste forms need to be used which will allow the radioactivity to decay away while the materials retain their integrity for thousands of years. The materials being used can be broken down into a few classes: glass waste forms, ceramic waste forms, and nanostructured materials.

The glass forms include borosilicate glasses and phosphate glasses. Borosilicate nuclear waste glasses are used on an industrial scale to immobilize high level radioactive waste in many countries which are producers of nuclear energy or have nuclear weaponry. The glass waste forms have the advantage of being able to accommodate a wide variety of waste-stream compositions, they are easy to scale up to industrial processing, and they are stable against thermal, radiative, and chemical perturbations. These glasses function by binding radioactive elements to nonradioactive glass-forming elements. Phosphate glasses while not being used industrially have much lower dissolution rates than borosilicate glasses, which make them a more favorable option. However, no single phosphate material has the ability to accommodate all of the radioactive products so phosphate storage requires more reprocessing to separate the waste into distinct fractions. Both glasses have to be processed at elevated temperatures making them unusable for some of the more volatile radiotoxic elements.

The ceramic waste forms offer higher waste loadings than the glass options because ceramics have crystalline structure. Also, mineral analogues of the ceramic waste forms provide evidence for long term durability. Due to this fact and the fact that they can be processed at lower temperatures, ceramics are often considered the next generation in high level radioactive waste forms. Ceramic waste forms offer great potential, but a lot of research remains to be done.

National Management Plans

Finland, the United States and Sweden are the most advanced in developing a deep repository for high-level radioactive waste disposal. Countries vary in their plans on disposing used fuel directly or after reprocessing, with France and Japan having an extensive commitment to reprocessing. The country-specific status of high-level waste management plans are described below.

In many European countries (e.g., Britain, Finland, the Netherlands, Sweden and Switzerland) the risk or dose limit for a member of the public exposed to radiation from a future high-level nuclear waste facility is considerably more stringent than that suggested by the International Commission on Radiation Protection or proposed in the United States. European limits are often more stringent than the standard suggested in 1990 by the International Commission on Radiation Protection by a factor of 20, and more stringent by a factor of ten than the standard proposed by the U.S. Environmental Protection Agency (EPA) for Yucca Mountain nuclear waste repository for the first 10,000 years after closure. Moreover, the U.S. EPA's proposed standard for greater than 10,000 years is 250 times more permissive than the European limit.

The countries that have made the most progress towards a repository for high-level radioactive waste have typically started with public consultations and made voluntary siting a necessary condition. This consensus seeking approach is believed to have a greater chance of success than top-down modes of decision making, but the process is necessarily slow, and there is "inadequate experience around the world to know if it will succeed in all existing and aspiring nuclear nations".

Moreover, most communities do not want to host a nuclear waste repository as they are "concerned about their community becoming a de facto site for waste for thousands of years, the health and environmental consequences of an accident, and lower property values".

References

- Waste-management-meaning-types-methods-importance-benefits-and-solutions, environmental-issues: indiacelebrating.com, Retrieved 8 April, 2019
- Cowan, G. A. (1976). "Oklo, A Natural Fission Reactor". Scientific American. 235 (1): 36. Bibcode:1976sciam.235a.36C. Doi:10.1038/scientificamerican0776-36. ISSN 0036-8733
- Solid-waste-management, technology: britannica.com, Retrieved 9 May, 2019
- "Oklo, Natural Nuclear Reactors". U.S. Department of Energy Office of Civilian Radioactive Waste Management, Yucca Mountain Project, DOE/YMP-0010. November 2004. Archived from the original on August 25, 2009. Retrieved September 15, 2009
- Main-types-of-solid-waste-management, solid-waste: yourarticlelibrary.com, Retrieved 10 June, 2019
- Organic-waste-management, greenhouse-best-management-practices-bmp-manual, greenhouse-floriculture: ag.umass.edu, Retrieved 11 July, 2019

3

Principles of Waste Management

The fundamental principles of waste management include waste minimization, resource efficiency, polluter pays principle, waste hierarchy, precycling, energy recovery and extended producer responsibility. All these diverse principles of waste management have been carefully analyzed in this chapter.

Sustainable waste management can be actively addressed through the planning process in the following ways:

- Reducing the quantity of materials required for the building.
- Reducing the amount of waste generated.
- Management of construction and demolition wastes.
- Materials specifications (e.g. use of reclaimed and recycled materials).
- Provision of recycling space/facilities.

The nationally accepted framework or approach for achieving reductions in waste arisings and sustainable waste management is the Waste Hierarchy.

Waste Hierarchy

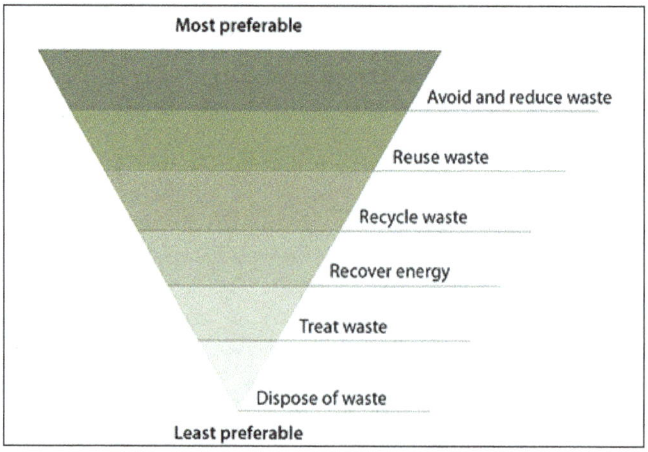

Basic principles of waste hierarchy.

The waste hierarchy provides a framework where waste management options are set out in priority

order to enable to correct choice to be made when assessing how to deal with waste. In the hierarchy, waste prevention is the most environmentally friendly and disposal the least.

Waste Reduction

By preventing waste before it occurs, money can be saved on the collection, treatment or disposals costs of waste. It also reduced the environmental impact and costs of extracting more raw materials, production and use.

This becomes more important when the true costs of waste are considered. The true cost of waste is not only the Disposal cost it also includes:

- Cost of purchasing materials/resources.
- Handling/processing/maintenance costs.
- Management time.
- Lost revenue.
- Any potential liabilities.

A study in 2009 showed that a typical construction skip costs around £1343 when you add the cost of the skip to the cost of labour and materials that fill it. The breakdown of this was:

- Skip hire £85 (quite low compared to current prices) – 6.4% of cost.
- Labour to fill it £163 – 12.1% of cost.
- Cost of materials in skip £1095 – 81.5% of cost.

It was estimated that a typical 80 m^2 house development would produce 5 skips of waste material, therefore, the financial cost of waste for a generic house construction was £6715, over 80% of which could be avoided through better resource use and selection of materials.

Materials Reuse

Reusing products and materials for the same (or alternative) purpose is the next preference. Before a material can be reused it should be assessed for its quality as it may be necessary to make minor repairs or additions before the product can reach the required standard.

This may include for example, retaining unused materials for one construction project to use on the next project.

Recycling and Composting

Recycling involves the collection, separation and processing of wastes to make new products, e.g. newspapers are regularly recycled either to make new newspapers or eco-friendly home insulation. Composting is the same process but with organic wastes, e.g. food waste composted to make new fertiliser products.

Recycling and composting processes usually require some energy to work well; however, the energy and cost to alternatively make new products from scratch are usually much greater. The economic viability of recycling/composting depends on factors such as the quality of the waste stream, the transport distances involved and the market price for the recycled materials which can fluctuate significantly. The aim should be to recycle construction wastes as close to their source as possible as they are typically heavy and bulky to transport.

Energy Recovery

Energy from waste incineration recovers a proportion of energy from the waste stream; however, usually much less than by recycling/composting, reusing or reducing the waste generated in the first instance.

Landfill Disposal

Disposal is the last option in the waste hierarchy and therefore the aim is to divert waste from this end destination. The only landfill in Hertfordshire currently accepting waste (Westmill in Ware) only has permission to continue recieving waste until 2017.

Proximity Principle

The transportation of waste can incur significant environmental and nuisance impacts plus unwanted additional cost. Therefore, the proximity principle encourages processing, recycling, reuse or disposal of waste as near to the point of its production as possible.

Benefits of Sustainable Waste Management

Sustainable waste management delivers lots of benefits:

- Reduced waste disposal costs (notably Landfill and Aggregates taxes).
- Reduced pressures on finite resources, such as virgin aggregates.
- Reduced greenhouse gas emissions from landfill and incineration.
- Reduced energy consumption from the manufacturing process.
- Increased economic productivity.
- Reduced requirement for additional landfill capacity.
- Reduced nuisance created by odour and visual intrusion from landfill sites.
- Improved corporate reporting and green credentials for business.

WASTE MINIMIZATION

Waste minimisation is a set of processes and practices intended to reduce the amount of waste produced. By reducing or eliminating the generation of harmful and persistent wastes, waste minimisation supports efforts to promote a more sustainable society. Waste minimisation involves redesigning products and processes and changing societal patterns of consumption and production.

The most environmentally resourceful, economically efficient, and cost effective way to manage waste often is to not have to address the problem in the first place. Managers see waste minimisation as a primary focus for most waste management strategies. Proper waste treatment and disposal can require a significant amount of time and resources; therefore, the benefits of waste minimisation can be considerable if carried out in an effective, safe and sustainable manner.

Traditional waste management focuses on processing waste after it is created, concentrating on re-use, recycling, and waste-to-energy conversion. Waste minimisation involves efforts to avoid creating the waste during manufacturing. To effectively implement waste minimisation the manager requires knowledge of the production process, cradle-to-grave analysis (the tracking of materials from their extraction to their return to earth) and details of the composition of the waste.

The main sources of waste vary from country to country. In the UK, most waste comes from the construction and demolition of buildings, followed by mining and quarrying, industry and commerce. Household waste constitutes a relatively small proportion of all waste. Industrial waste is often tied to requirements in the supply chain. For example, a company handling a product may insist that it should be shipped using particular packing because it fits downstream needs.

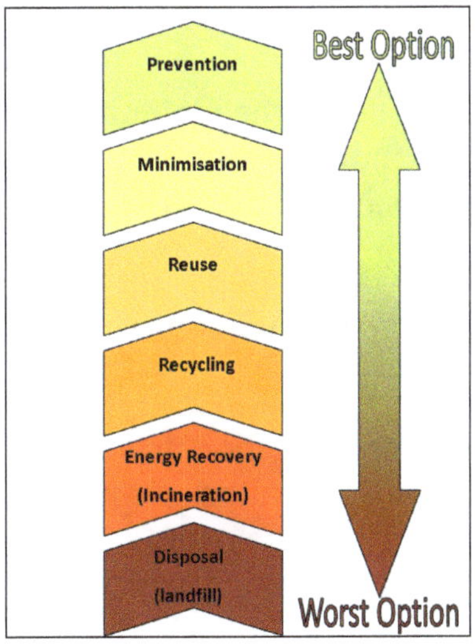

Waste hierarchy: Refusing, reducing, reusing, recycling and composting allow to reduce waste.

Benefits

Waste minimisation can protect the environment and often turns out to have positive economic benefits. Waste minimisation can improve:

- Efficient production practices: Waste minimisation can achieve more output of product per unit of input of raw materials.
- Economic returns: More efficient use of products means reduced costs of purchasing new materials improving the financial performance of a company.

- Public image: The environmental profile of a company is an important part of its overall reputation and waste minimisation reflects a proactive movement towards environmental protection.
- Quality of products produced: New innovation and technological practices can reduce waste generation and improve the quality of the inputs in the production phase.
- Environmental responsibility: Minimising or eliminating waste generation makes it easier to meet targets of environmental regulations, policies, and standards. The environmental impact of waste will be reduced.

Industries

In industry, using more efficient manufacturing processes and better materials generally reduces the production of waste. The application of waste minimisation techniques has led to the development of innovative and commercially successful replacement products.

Waste minimisation efforts often require investment, which is usually compensated by the savings. However, waste reduction in one part of the production process may create waste production in another part.

Processes

- Reuse of scrap material

Scraps can be immediately re-incorporated at the beginning of the manufacturing line so that they do not become a waste product. Many industries routinely do this; for example, paper mills return any damaged rolls to the beginning of the production line, and in the manufacture of plastic items, off-cuts and scrap are re-incorporated into new products.

- Improved quality control and process monitoring

Steps can be taken to ensure that the number of reject batches is kept to a minimum. This is achieved by increasing the frequency of inspection and the number of points of inspection. For example, installing automated continuous monitoring equipment can help to identify production problems at an early stage.

- Waste exchanges

This is where the waste product of one process becomes the raw material for a second process. Waste exchanges represent another way of reducing waste disposal volumes for waste that cannot be eliminated.

- Ship to point of use

This involves making deliveries of incoming raw materials or components direct to the point where they are assembled or used in the manufacturing process to minimise handling and the use of protective wrappings or enclosures.

- Zero waste

This is a whole systems approach that aims to eliminate waste at the source and at all points down the supply chain, with the intention of producing no waste. It is a design philosophy which emphasizes waste prevention as opposed to end of pipe waste management. Since, globally speaking, waste as such, however minimal, can never be prevented (there will always be an end-of-life even for recycled products and materials), a related goal is pollution prevention.

Product Design

Waste minimisation and resource maximisation for manufactured products can most easily be done at the design stage. Reducing the number of components used in a product or making the product easier to take apart can make it easier to be repaired or recycled at the end of its useful life.

In some cases, it may be best not to minimise the volume of raw materials used to make a product, but instead reduce the volume or toxicity of the waste created at the end of a product's life, or the environmental impact of the product's use.

Fitting the Intended Use

In this strategy, products and packages are optimally designed to meet their intended use. This applies especially to packaging materials, which should only be as durable as necessary to serve their intended purpose. On the other hand, it could be more wasteful if food, which has consumed resources and energy in its production, is damaged and spoiled because of extreme measures to reduce the use of paper, metals, glass and plastics in its packaging.

Durability

Improving product durability, such as extending a vacuum cleaner's useful life to 15 years instead of 12, can reduce waste and usually much improves resource optimisation.

But in some cases it has a negative environmental impact. If a product is too durable, its replacement with more efficient technology is likely to be delayed. Therefore, extending an older machine's useful life may place a heavier burden on the environment than scrapping it, recycling its metal and buying a new model. Similarly, older vehicles consume more fuel and produce more emissions than their modern counterparts.

Most proponents of waste minimisation consider that the way forward may be to view any manufactured product at the end of its useful life as a resource for recycling and reuse rather than waste.

Making refillable glass bottles strong enough to withstand several journeys between the consumer and the bottling plant requires making them thicker and so heavier, which increases the resources required to transport them. Since transport has a large environmental impact, careful evaluation is required of the number of return journeys bottles make. If a refillable bottle is thrown away after being refilled only several times, the resources wasted may be greater than if the bottle had been designed for a single journey. Many choices involve trade-offs of environmental impact, and often there is insufficient information to make informed decisions.

Retail

Various aspects of business practices affect waste, such as the use of disposable tableware in restaurants.

Reusable Shopping Bags

Reusable bags are a visible form of re-use, and some stores offer a "bag credit" for re-usable shopping bags, although at least one chain reversed its policy, claiming "it was just a temporary bonus". In contrast, one study suggests that a bag tax is a more effective incentive than a similar discount. Of note, the before/after study compared a circumstance in which some stores offered a discount vs. a circumstance in which all stores applying the tax. While there is a minor inconvenience involved, this may remedy itself, as reusable bags are generally more convenient for carrying groceries.

Households

Appropriate amounts and sizes can be chosen when purchasing goods; buying large containers of paint for a small decorating job or buying larger amounts of food than can be consumed create unnecessary waste. Also, if a pack or can is to be thrown away, any remaining contents must be removed before the container can be recycled.

Home composting, the practice of turning kitchen and garden waste into compost can be considered waste minimisation.

The resources that households use can be reduced considerably by using electricity thoughtfully (e.g. turning off lights and equipment when it is not needed) and by reducing the number of car journeys made. Individuals can reduce the amount of waste they create by buying fewer products and by buying products which last longer. Mending broken or worn items of clothing or equipment also contributes to minimising household waste. Individuals can minimise their water usage, and walk or cycle to their destination rather than using their car to save fuel and cut down emissions.

In a domestic situation, the potential for minimisation is often dictated by lifestyle. Some people may view it as wasteful to purchase new products solely to follow fashion trends when the older products are still usable. Adults working full-time have little free time, and so may have to purchase more convenient foods that require little preparation, or prefer disposable nappies if there is a baby in the family.

The amount of waste an individual produces is a small portion of all waste produced by society, and personal waste reduction can only make a small impact on overall waste volumes. Yet, influence on policy can be exerted in other areas. Increased consumer awareness of the impact and power of certain purchasing decisions allows industry and individuals to change the total resource consumption. Consumers can influence manufacturers and distributors by avoiding buying products that do not have eco-labelling, which is currently not mandatory, or choosing products that minimise the use of packaging. In the UK, PullApart combines both environmental and consumer packaging surveys, in a curbside packaging recycling classification system to minimise waste. Where reuse schemes are available, consumers can be proactive and use them.

Health-care Facilities

Health-care establishments are massive producers of waste. The major sources of health-care waste are: hospitals, laboratories and research centres, mortuary and autopsy centres, animal research and testing laboratories, blood banks and collection services, and nursing homes for the elderly.

Waste minimisation can offer many opportunities to these establishments to use fewer resources, be less wasteful and generate less hazardous waste. Good management and control practices among health-care facilities can have a significant effect on the reduction of waste generated each day.

Practices

There are many examples of more efficient practices that can encourage waste minimization in healthcare establishments and research facilities.

Source Reduction

- Purchasing reductions which ensures the selection of supplies that are less wasteful or less hazardous.
- The use of physical rather than chemical cleaning methods such as steam disinfection instead of chemical disinfection.
- Preventing the unnecessary wastage of products in nursing and cleaning activities.

Management and Control Measures at Hospital Level

- Centralized purchasing of hazardous chemicals.
- Monitoring the flow of chemicals within the health care facility from receipt as a raw material to disposal as a hazardous waste.
- The careful separation of waste matter to help minimize the quantities of hazardous waste and disposal.

Stock Management of Chemical and Pharmaceutical Products

- Frequent ordering of relatively small quantities rather than large quantities at one time.
- Using the oldest batch of a product first to avoid expiration dates and unnecessary waste.
- Using all the contents of a container containing hazardous waste.
- Checking the expiry date of all products at the time of delivery.

WASTE HIERARCHY

The waste hierarchy, or the hierarchy of waste management is a series of five or six options for dealing with waste. It is a more complex version of the "reduce - reuse - recycle" green arrowed triangle logo which ranks as how favourable each option is from an environmental perspective. However, it tends not to say anything about cost or practicality.

The Hierarchy

The standard outline of the hierarchy ranks 6 approaches to waste management. In order (most preferable to least preferable) these are: prevention, minimisation, reuse, recycle, energy recovery and disposal.

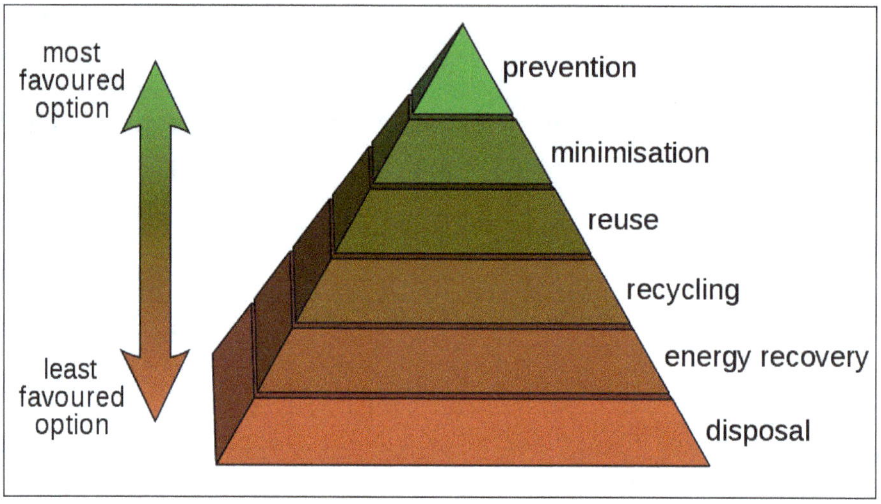

It has an unusual truth in advertising, in that the width of each step on the pyramid approximates how much that step is actually done in practice. This is important as people tend to believe that recycling is the be-all and end-all of environmental protection - it isn't, as the steps below outline.

Prevention

The top level is "prevention" or "avoidance". This waste is prevented in the most obvious way: by avoiding its production entirely. For example, we can prevent plastic carrier bags from going to a landfill by using long-life bags, such as canvas bags, which last virtually forever. In green chemistry, completely different routes for chemical synthesis are developed, with new reagents that cut down on waste and improve efficiency. Alternatively, we just completely avoid the source entirely by just not making the end product, like disposable novelty gadgets that don't do anything and just get thrown away anyway.

Often in industry, this is the least practical and most expensive because of the amount of research and development needed to completely avoid or replace a substance or process. Completely abandoning cars or any form of transport, for example, is practically impossible. In these cases we must look to making the processes as efficient as possible, which brings us to the next step.

Minimisation

Reducing waste is self-explanatory, and relies on making a process more efficient. This option often applies to industrial processes far more than the common domestic environment. People may think it's odd when chemical engineers tinker with a chemical reaction to change its yield from 99.5% to 99.75%

(an apparently minuscule increase) but this halves the amount of waste side products produced. Generally, improved efficiency is synonymous with economic viability as well, as more useful output can be gained from the same input, and less waste is less expensive to dispose of.

This can also apply outside industry, by homes using more efficient appliances or using their electric supply more effectively, from not leaving heating on constantly or water running too long, to boiling a kettle with only as much water as you want to use.

Reuse

This is simple; reduce waste by not wasting something, but reusing it. This can be combined with the top level by replacing something disposable with something reusable. But most commonly, it means to quite literally reuse something. Plastic bags being, again, the most obvious example. Do not throw these things away. Polyethylene bags (at least the ones that haven't had silly degradation additives put into them) are extremely strong and durable for their weight and thickness. You can reuse these dozens upon dozens of times before they break and avoid accumulating hundreds of them, or use them as bin bags rather than buying new bits of plastic.

Creative Reuse or Up-cycling is the reuse of unwanted items by re-purposing them into new end use items.

A further extension to the reuse idea is refilling. Ink cartridges are an example, in which the only bit replaced is the stuff that is used up (i.e., the ink). Glass milk bottles are also often collected, washed, refilled and delivered many times in their life. Reuse is often measured in terms of the amount of material diverted from landfills, or the value of the material retained.

Recycle

It often surprises people how far down the hierarchy recycling is. Despite it being a very visual and pleasing way of waste management (it makes people feel like they're doing something) it's not incredibly efficient as a general process. Recycling is defined as putting energy into a waste item to convert it to something else entirely, sometimes with lower grade and value. It is the preferred option when reuse is not an option, if an item is broken, or in a poor condition that means it can't be reused. Recycling contrasts with reuse in the fact that recycling tends to break an item down to its raw materials before making them into something else. In the milk bottle example, reusing involves washing and refilling, while recycling would involve grinding the glass and melting it again, turning the raw material into a different grade of glass.

Recycling saves on processing or extracting of raw materials so is useful when materials are limited and reuse is difficult. For example, recycling aluminum cans uses less than 5% of the energy used to purify aluminum metal from bauxite, which requires very power-intensive electrolysis. With this step removed in the recycling process, the metal just needs to be melted down in order for it to be reused in any way. This makes it one of the few recycling processes that not only proves to be very efficient, but can also turn a profit.

However, there are numerous drawbacks to recycling. If the recycling process uses up more energy or material resources (and this is very often true of novelty recycling products) then it is not particularly viable and other options from higher in the list should be considered.

Downcycling

A large proportion of recycling is actually what is termed downcycling. This is where materials are recycled into a lower grade of material that will later be thrown away, such as plastic bottles being shredded to make low quality disposable plastic, or paper being recycled as toilet paper. This is less desirable than normal recycling and may even be less desirable than incineration depending on how much energy the downcycling process takes the disposal method for the second generation product. However, in places where only lower quality material is needed (toilet paper being the most prominent example) using downcycling to access the raw materials is preferable to using brand new materials.

Energy Recovery

"Energy recovery" is really a euphemism for "burn it". This may seem odd, but waste materials can be burned down and the energy from it harnessed as electricity. This saves burning oil or coal, although other oxidised pollutants, namely sulphur-containing products and dioxins produced can cause more damage. Burning waste in this manner needs to be tightly controlled.

Energy recovery can also be achieved with bio-mass waste. We can either burn it directly, or compost it to capture methane, which can be burned to produce energy (biogas).

Disposal

"Disposal" is also known as "burying it". This is the last resort method for waste disposal and the least favoured (in waste reduction terms), although still the most common. Mostly it is least favoured because we don't get anything back from what we put in the ground. Once something is in a landfill and buried (provided the landfill is set up and lined properly) there is no real harm the waste can do as it degrades naturally - although this can take thousands of years. Some CO_2 or other greenhouse gases may be released, however. Overall, sending something to a landfill should be the last resort when the above five options have been tried.

Unfortunately, it is the most common, mostly because of laziness or ignorance on behalf of the public to follow the above, or the lack of research in reuse and recycling things efficiently - hopefully, this should improve in the future, with added awareness and incentives for commercial operations to follow real waste management strategies, rather than simply greenwashing their image.

Many landfills also work towards "energy recovery" by salvaging methane and other gases to be used for energy and heating. This is vastly preferable to simply allowing methane to leak out, as it is a significant greenhouse gas. While methane will eventually decompose in the atmosphere to less harmful carbon dioxide, burning it for useful purposes accelerates the decomposition for a net benefit (or rather, net less-harmful), and produces energy that would otherwise have to be acquired by other means.

PRECYCLING

Precycling is the practice of reducing waste by attempting to avoid bringing items which will generate waste into home or business. The U.S. Environmental Protection Agency (EPA) also cites that precycling is the preferred method of integrated solid waste management because it cuts waste at its source and therefore trash is eliminated before it is created. According to the EPA, precycling is also characterized as a decision-making process on the behalf of the consumer because it involves making informed judgments regarding a product's waste implications. The implications that are taken into consideration by the consumer include: whether a product is reusable, durable, or repairable; made from renewable or non-renewable resources; over-packaged; and whether or not the container is reusable.

Precycling has the ability to build industrial, social, environmental, and economic circumstances that allow for old products to be converted into new resources:

- Industrial: Increasing the independence from accumulative substances, such as heavy metals, fossil fuels, synthetics, etc.
- Economic: Create a circular economy.
- Ecological/environmental: Allowance for more extensive and diverse natural habitats where the resources are returned to nature.
- Societal: Extend the capacity of precycling to meet everyone's needs.

The concept of 'precycling' was coined in 1988 by social marketing executive Maureen O'Rorke in a public waste education campaign for the City of Berkeley. The application of precycling is not limited to large corporations, but can be administered on smaller scales in local communities. The reason precycling is effective on large scales and on small scales stems from the idea that it shares a common language between experts and non-experts, buyers and sellers, economists and environmentalists. However, it is important to consider that waste prevention systems, such as precycling, require the collaborative effort from several working parts. These parts include prevention targets, producer responsibility, householder charging, funding for pilot projects, public involvement, engagement of private and third sectors, and public campaigns that spread awareness.

Integration of Waste Management

The original three-pronged push for waste management is "Reduce, Reuse, Recycle." Precycling emphasizes "reducing and reusing", while harnessing and questioning the momentum and popularity of the term "recycle." In addition to this strategy, precycling incorporates four supplementary R's: Repair, Recondition, Remanufacture and Refuse. Waste is a resource that can be reused, recycled, recovered, or treated. Precycling differs from other singular forms of waste prevention because it encompasses not one behavior, but many.

Reduce

Reduce is a form of precycling that allows for the preservation of natural resources and also saves money on behalf of the manufacturer, the consumer, and the waste manager. Moreover, effective source reduction slows the depletion of environmental resources, prolongs the life of waste management facilities, and makes combustion and landfills safer by removing toxic waste components.

Reuse

Reuse is a form of precycling that reinvents items after their initial life and avoids creating additional waste.

Recycle

Although precycling harnesses the familiarity of the term recycling, it is still important to note the difference between recycling and prevention. Since precycling focuses on the prevention of waste production, this entails that measures are taken before a substance, material, or product has become waste. Whereas recycling is a type of precycling that involves taking action before existing waste is abandoned to nature. Recycling is a process where discarded materials are collected, sorted, processed, and used in the production of new products. Every time a person engages in the act of recycling, they help increase the market and bring the cost down. However, current research from the American Plastics Council states that only 25% of the nation's recycling capabilities are being utilized.

Traditionally recycling requires large amounts of energy to "melt down" and then re-manufacture items. While this may cut down on the amount of trash that is going into landfills, it is not sustainable unless the underlying energy supply is sustainable. In addition, recycling often means downcycling and always involves at least some loss of the original material, so primary extraction is still required to make up the difference. Precycling reduces these problems by using less material in the first place, so less has to be recycled.

Repairing

Repair is a type of precycling that corrects specified faults in a product, however the quality of a repaired product is inferior to reconditioned or remanufactured items. One survey found that 68% of the respondents believed repairing was not cost efficient and sought alternative methods such as reconditioning or remanufacturing.

Reconditioning

Reconditioning is a type of precycling that requires the rebuilding of major components to restore a product's working condition, which is expected to be inferior to the original product.

Remanufacturing

Remanufacturing is another type of precycling that involves the greatest degree of work content, which results in superior quality of the product. In order to remanufacture a product, it requires a total dismantling of the product and the subsequent restoration and replacement of its parts. Remanufacturing is a preferred method of waste reduction compared to repairing and reconditioning because it preserves the embodied energy that has been used to shape the components of a product for their first life and it only requires 20-25% of the initial energy used in formation.

Refuse

Refusal to buy certain products due to detrimental impacts on the environment or wasteful packaging is another type of precycling because the rejection of such items paves the way for products that can be reduced, reused, or recycled.

Zero-waste Strategy

A zero waste approach aims to prevent rather than just reduce accumulated waste. Zero-waste goes beyond recycling to include the whole system, which includes the flow of resources and waste through human society. This "design principle" works to maximize recycling, minimize waste, reduce consumption and ensures that products are reused, repaired or recycled back into nature or the market. This preventative approach is more manageable and effective than incremental approaches that focus on gradually reducing the amount of impact because it is less complex and contains less information, which permits wider public participation.

Sustainability

In regards to sustainability, the term itself is often associated with resource constraints and maintenance of the status quo rather than growth and prosperity. However, with the implementation of a zero-waste management strategy, sustainable practices can push the status quo in order to create a society that is capable of development, technically and culturally advanced, dynamic in population and production, thoughtful with the use of non-renewable resources, and diverse, democratic, and challenging.

Economic Effects

Increased waste production is often negatively associated with increased economic growth. However, a zero-waste management strategy allows for economic growth that works cohesively with sustainability rather than against it. The implementation of a zero-waste strategy is part of an economic goal-set that aims to create a circular economy. A circular economy refers to a closed-loop socio-economic system that focuses on minimizing wastes while simultaneously maximizing stocks of resources for the economy. This closed-loop design diverts linear (open-loop) waste disposal streams into new raw material streams.

In a circular economy, one way to minimize waste is through the employment of precycling insurance, which allows for a full range of financed waste prevention opportunities. This type of insurance would set premiums related to the risk of a product ending up as waste, and these premiums would serve to fund actions concerning waste prevention. When establishing a premium for precycling insurance several factors need to calculated: recyclability or biodegradability; provision of infrastructure, habitat or collaborations for the generation of the product from new resources; the ecosystem concentrations of product components above natural levels. The idea of precycling insurance is plausible considering the aim of insurance industries is to avoid losses rather than paying for losses. However, in order for this idea to work, private and third sectors need to be involved and engaged in the issue. In this instance, a third sector refers to small charities and a handful of societal enterprises that coordinate with charity shops.

Environmental Effects

According to the "Extended Producer Responsibility" principle, impacts are substantially determined at the point of design where key decisions are made on materials, production process, and how products are used and disposed of at the end of life-cycle, which falls on the producer. However, in a circular economy there is the recognition that nature's capacity needs to be maximized through the reprocess of biodegradable wastes produced by industries and human activity. This task is accomplished through the procurement and funding of precycling insurance premiums that invest in systematic preservation of endangered habitats, careful harvesting of biological resources and expansions of productive ecosystems. Additionally, in terms of climate change, precycling insurance offers a flexible alternative to the binding limits on greenhouse gas emissions and international taxation on mineral fuels. In terms of waste management systems, the environment benefits from the reparation of products to the greatest degree because less energy is required and the majority of the original material is kept intact.

Societal Effects

The social structure operating under a circular economy is referred to as a circular society. The aim of a circular society is to create a cooperative culture by means of problem-prevention, resource-availability and fuller participation, with reference to precycling. One critique of this approach, in terms of waste management, is that it is difficult to maintain a cooperative culture within a society because it is constantly evolving and changing.

Raising Awareness

There is an increasing public awareness on the need for sustainable production and consumption. One campaign that aimed at raising awareness of precycling focused on whether people's self-reported behaviors were affected by exposure to precycling advertisements on the radio, television, or in-store flyers. The researchers concluded that the most effective results stemmed from the inclusion of social rewards that invoke an intrinsic motivation to engage in precycling behaviors.

Another way to raise awareness is through statistics that highlight the potential impacts that can be achieved through waste prevention. For instance, if 70 million Americans bought a half-gallon plastic-coated carton container of milk each week (instead of two quarts), then 41.6 million pounds of paper discards and 5.7 pounds of plastic discards would be reduced annually. This transition from two quarts to a half-gallon would save $145.6 million on packaging each year.

Implementation

In order to effectively implement precycling practices and behaviors, the public needs to feel "enabled", "engaged", "encouraged", and "exemplified" in their efforts to partake in precycling.

Not only can the average consumer practice precycling, but industries can also participate. Purchasing from parts suppliers, reuse of chemicals, and reduction of unnecessary packaging are some methods. There are some companies and countries that have taken it upon themselves to implement more sustainable practices that align with precycling principles. For instance, Fonterra reduced its packaging through the implementation of bulking, reuse and redesign. Further, Waste Management New Zealand created Recycle New Zealand, which provided a subsidiary focusing on the collection of materials that could be diverted and sorted prior to the operations of reducing, recycling, or recovering. Moving forward, free-trade organizations can further implement precycling practices by exploring this strategy as a new way to reduce regulations and to promote greater industrial freedom of choice.

Moreover, the individual consumer can develop precycling habits by engaging in the following practices and behaviors:

Enviro-shopping

Enviro-shopping is considered shopping with the environment and implements a precycling strategy:

- Bringing one's own grocery bag or bring old ones back to the store.
- Buying packages with the least amount of packaging.
- Buying in bulk, but not buying more than one will use.
- Looking for products with reusable dishes.

Product Selection

Products to choose from in accordance with precycling principles:

- Plastic milk jugs or glass milk containers (no cartons).
- Fresh fruit and vegetables.
- Concentrated products that involves less packaging.
- Recycled products.
- Rechargeable batteries.

Behaviors

In addition to shopping practices that implement precycling principles, there are also behaviors that can be undertaken to prevent waste:

- Home composting.
- Avoid junk mail.
- Buy second-hand.

One way to participate in precycling is to carry a "precycling kit". Include a Tupperware or non-disposable container, silverware set, a cloth napkin or handkerchief, and a thermos or water-bottle within a cloth bag that can double as a grocery/shopping bag.

ZERO WASTE

Zero Waste is a set of principals focused on waste prevention that encourages the redesign of resource life cycles so that all products are reused. The goal is for no trash to be sent to landfills, incinerators, or the ocean. Currently, only 9% of plastic is actually recycled. The process recommended is one similar to the way that resources are reused in nature. The definition adopted by the Zero Waste International Alliance (ZWIA) is:

> Zero Waste: The conservation of all resources by means of responsible production, consumption, reuse, and recovery of all products, packaging, and materials, without burning them, and without discharges to land, water, or air that threaten the environment or human health.

Zero Waste refers to waste prevention as opposed to end-of-pipe waste management. It is a whole systems approach that aims for a massive change in the way materials flow through society, resulting in no waste. Zero waste encompasses more than eliminating waste through recycling and reuse, it focuses on restructuring production and distribution systems to reduce waste. Zero waste is more of a goal or ideal rather than a hard target. Zero Waste provides guiding principles for continually working towards eliminating wastes.

Advocates expect that government regulation is needed to influence industrial choices over product and packaging design, manufacturing processes, and material selection. Advocates say eliminating waste decreases pollution, and can also reduce costs due to the reduced need for raw materials.

Used products dumped in a landfill.

Cradle-to-cradle or Cradle-to-grave

Cradle-to-grave is a term used to describe a linear model for materials that begins with resource extraction, moves to product manufacturing, and, ends by a 'grave', where the product is disposed of in a landfill. Cradle-to-grave is in direct contrast to cradle-to-cradle. Cradle-to-cradle is a term used in life-cycle analysis to describe a material or product that is recycled into a new product at the end of its life, so that ultimately there is no waste.

Cradle-to-cradle focuses on designing industrial systems so that materials flow in closed-loop cycles which mean that waste is minimized, and waste products can be recycled and reused. Cradle-to-cradle simply goes beyond dealing with issues of waste after it has been created, by addressing problems at the source and by re-defining problems by focusing on design. The cradle-to-cradle model is sustainable and considerate of life and future generations.

The cradle-to-cradle framework has evolved steadily from theory to practice. In the industrial sector, it is creating a new notion of materials and material flows. Just as in the natural world, in which one organism's 'waste', cycles through an ecosystem to provide nourishment for other living things, cradle-to-cradle materials circulate in closed-loop cycles, providing nutrients for nature or industry.

An example of a closed loop, cradle-to-cradle product design is DesignTex Fabric. It has designed an upholstery fabric, Climatex Lifecycle, which is a blend of pesticide- and residue-free wool and organically grown ramie, dyed and processed entirely with nontoxic chemicals.

The spread of industrialization worldwide has been accompanied by a large increase in waste production. In 2012 the World Bank stated that 1.3billion tonnes of municipal waste was produced by urban populations and estimates that that number will reach 2.2 billion tonnes by 2025 (Global Solid Waste Management Market - Analysis and Forecast). The increase in solid waste production increases the need for landfills. With the increase in urbanization, these landfills are being placed closer to communities. These landfills are disproportionately located in areas of low socioeconomic status with primarily non-white populations. Findings indicated these areas are often targeted as waste sites because permits are more easily acquired and there was generally less community resistance. Additionally, within the last five years, more than 400 hazardous waste facilities have received formal enforcement actions for unspecified violations that were considered to be a risk to human health.

There is a growing global population that is faced with limited resources from the environment. To relieve the pressures placed on the finite resources available it has become more important to prevent waste. To achieve zero waste, waste management has to move from a linear system to be more cyclical so that materials, products, and substances are used as efficiently as possible. Materials must be chosen so that it may either return safely to a cycle within the environment or remain viable in the industrial cycle.

Zero waste promotes not only reuse and recycling but, more importantly, it promotes prevention and product designs that consider the entire product life cycle. Zero waste designs strive for reduced materials use, use of recycled materials, use of more benign materials, longer product lives, reparability, and ease of disassembly at end of life. Zero waste strongly supports sustainability by protecting the environment, reducing costs and producing additional jobs in the management and handling of wastes back into the industrial cycle. A zero waste strategy may be applied to businesses, communities, industrial sectors, schools and homes.

Benefits proposed by advocates include:

- Saving money: Since waste is a sign of inefficiency, the reduction of waste can reduce costs.
- Faster Progress: A zero waste strategy improves upon production processes and improving environmental prevention strategies which can lead to take larger, more innovative steps.
- Supports sustainability: A zero waste strategy supports all three of the generally accepted goals of sustainability - economic well-being, environmental protection, and social well-being.
- Improved material flows: A zero waste strategy would use far fewer new raw materials and send no waste materials to landfills. Any material waste would either return as reusable or recycled materials or would be suitable for use as compost.

Health

A major issue with landfills is hydrogen sulfide, which is released during the natural decay of waste. Studies have shown a positive association between increased lung cancer mortality rates and increased morbidity and mortality related to respiratory disease and hydrogen sulfide exposure. These studies also showed that the hydrogen sulfide exposure increased with proximity to the landfill.

Household chemicals and prescription drugs are increasingly being found in large quantities in the leachate from landfills. This is causing concern about the ability of landfills to contain these materials and the possibility of these chemicals and drugs making their way into the groundwater and the surrounding environment.

Zero waste promotes a circular material flow that allows materials to be used over and over, reducing the need for landfill space. Through zero waste the number of toxins released into the air and water would be decreased and products examined to determine what chemicals are used in the production process.

Health Issues related to landfills:

- Birth Defects and Low birth weight: Associated with close proximity to landfills, exposure to particle matter and Nitrogen dioxide.
- Respiratory disease and Lung Cancer: Related to the release of Hydrogen Sulfide.

Zero waste's promotion of a cyclical product life can help reduce the need to create and fill landfills. This can help reduce incidences of respiratory diseases and birth defects that are associated with the toxins released from landfills. Zero waste also can help preserve local environments and drinking water sources by preventing pollutants from entering the ecosystem.

Packaging Example

Returnable glass milk bottles.

Milk can be shipped in many forms. One of the traditional forms was reusable returnable glass milk bottles, often home delivered by a milkman. While some of this continues, other options have recently been more common: one-way gable-top paperboard cartons, one-way aseptic cartons, one-way recyclable glass bottles, one-way milk bags, and others. Each system claims some advantages and also has possible disadvantages. From the zero waste standpoint, the reuse of bottles is beneficial because the material usage per trip can be less than other systems. The primary input (or resource) is silica-sand, which is formed into glass and then into a bottle. The bottle is filled with milk and distributed to the consumer. A reverse logistics system returns the bottles for cleaning, inspection, sanitization, and reuse. Eventually, the heavy-duty bottle would not be suited for further use and would be recycled. Waste and landfill usage would be minimized. The material waste is primarily the wash water, detergent, transportation, heat, bottle caps, etc. While true zero waste is never achieved, a life cycle assessment can be used to calculate the waste at each phase of each cycle.

Recycling and Rotting

It is important to distinguish recycling from zero waste. Some claim that the key component to zero waste is recycling while others reject that notion in favor of reusing high function. The common understanding of recycling is simply that of placing bottles and cans in a recycle bin. The modern version of recycling is more complicated and involves many more elements of financing and government support. For example, a 2007 report by the U.S. Environmental Protection Agency states that the US recycles at a national rate of 33.4% and includes in this figure composted materials. In addition many worldwide commodity industries have been created to handle the materials that are recycled. At the same time, claims of recycling rates have sometimes been exaggerated, for example by the inclusion of soil and organic matter used to cover garbage dumps daily, in the "recycled" column. In states with recycling incentives, there is constant local pressure to pump up the recycling rate figures.

The movement toward recycling has separated itself from the concept of zero waste. One example of this is the computer industry where worldwide millions of PC's are disposed of as electronic waste each year (160 million in 2007). Those computers that enter the recycling stream are broken down into a small amount of raw materials while most merely enter dumps through export to third world countries. Companies are then able to purchase some raw materials, notably steel, copper and glass, reducing the use of new materials. On the other hand, there is an industry, more aligned with the Zero Waste principle of design for long term reuse, that actually repairs computers. It is called the Computer Refurbishing industry and it predates the current campaign to just collect and ship electronics. They have organizations and conferences and have for many years donated computers to schools, clinics and non-profits. Zero Waste planning demands that components be redesigned for effective reuse over long lives leading to even more refurbishing and repair.

There is one seminal example that brings out the difference between Zero Waste and recycling in stark relief. That example, quoted in Getting To Zero Waste, is the software business. Zero Waste is sensitive to the waste of intellectual effort that would be caused by the need to recreate certain basic inventions of software (called objects in software design) as opposed to copying them over and over whenever needed. The waste would occur as the software developers consume resources while solving problems already solved earlier. The application of Zero Waste analysis is straightforward as it recommends conserving human effort. On the other hand, the usual approach of recycling would be to look for some materials that could be found to reuse. The materials on which software is saved (such as paper or diskettes) is of little significance compared to the saving of human effort and if software is saved electronically, there is no media at all. Thus Zero Waste correctly identifies a wasteful behavior to avoid while recycling has no application.

The recycling movement has been embraced by the garbage industry because it serves so well as greenwashing i.e. a way to show that design for garbage creation is acceptable because materials will be kept out of a dump by recycling them. Zero Waste, on the other hand, offers the garbage industry no such screen against public condemnation of waste, and therefore actually threatens the continued need for garbage disposal. For example, in Alameda County, California, garbage dumping is charged a surcharge of $8/ton (as of 2009) which goes entirely for a recycling subsidy but none of which goes for any kind of Zero Waste style designing. Zero Waste has received no support from the garbage industry or politicians under their control except in those cases where it can be claimed to consist solely of more recycling.

Rethink, Refuse, Reduce, Reuse and Recycle

Zero waste is poorly supported by the enactment of government laws to enforce the waste hierarchy of refuse, reduce, reuse, recycle and rot (compost). In practice, these laws invariably emphasize destruction and recycling, while the reuse component is marginalized.

A special feature of Zero Waste as a design principle is that it can be applied to any product or process, in any situation or at any level. Thus it applies equally to toxic chemicals as to benign plant matter. It applies to the waste of atmospheric purity by coal-burning or the waste of radioactive resources by attempting to designate the excesses of nuclear power plants as "nuclear waste". All processes can be designed to minimize the need for discard, both in their own operations and in the usage or consumption patterns which the design of their products leads to. Recycling, on the other hand, deals only with simple materials.

Zero Waste can even be applied to the waste of human potential by enforced poverty and the denial of educational opportunity. It encompasses redesign for reduced energy wasting in industry or transportation and the wasting of the earth's rainforests. It is a general principle of designing for the efficient use of all resources, however defined.

The recycling movement may be slowly branching out from its solid waste management base to include issues that are similar to the community sustainability movement.

Zero waste, on the other hand, is not based in waste management limitations to begin with but requires that we maximize our existing reuse efforts while creating and applying new methods that minimize and eliminate destructive methods like incineration and recycling. Zero Waste strives to ensure that products are designed to be repaired, refurbished, re-manufactured and generally reused.

Online web services, like Free Cycle or the reGives Network have risen in popularity over the last decade where locals can give items that they no longer need to others locally in an effort to keep items out of landfills and work toward a zero waste lifestyle.

Significance of Dump Capacity

Many dumps are currently exceeding carrying capacity. This is often used as a justification for moving to Zero Waste. Others counter by pointing out that there are huge tracts of land available throughout the USA and other countries which could be used for dumps. Proposals abound to destroy all garbage as a way to solve the garbage problem. These proposals typically claim to convert all or a large portion of existing garbage into oil and sometimes claim to produce so much oil that the world will henceforth have abundant liquid fuels. One such plan, called Anything Into Oil, was promoted by Discover Magazine and Fortune Magazine in 2004 and claimed to be able to convert a refrigerator into "light Texas crude" by the application of high-pressure steam.

Corporate Initiatives

An example of a company that has demonstrated a change in landfill waste policy is General Motors (GM). GM has confirmed their plans to make approximately half of its 181 plants worldwide "landfill-free" by the end of 2010. Companies like Subaru, Toyota, and Xerox are also producing landfill-free plants. Furthermore, The United States Environmental Protection Agency (EPA) has worked with GM and other companies for decades to minimize the waste through its WasteWise program. The goal for General Motors is finding ways to recycle or reuse more than 90% of materials by: selling scrap materials, adopting reusable parts boxes to replace cardboard, and even recycling used work gloves. The remainder of the scraps might be incinerated to create energy for the plants. Besides being nature-friendly, it also saves money by cutting out waste and producing a more efficient production. All these organizations push forth to make our world clean and producing zero waste.

A garden centre in Faversham, UK, has started to prevent plastic plant pots from being passed down to customers. Instead, it reuses the plastic pots only locally in the garden center, but upon selling it to its customers it repots the plants in paper plant pots. It also sells plants wrapped in hessian, and uses a variety of techniques to prevent handing down (single-use) plastics to customers.

Rot of Waste

The waste sent to landfills may be harvested as useful materials, such as in the production of solar energy or natural fertilizer/de-composted manure for crops.

It may also be reused and recycled for something that we can actually use. "The success of General Motors in creating zero-landfill facilities shows that zero-waste goals can be a powerful impetus for manufacturers to reduce their waste and carbon footprint," says Latisha Petteway, a spokesperson for the EPA.

Construction and Deconstruction

Zero Waste is a goal, a process, a way of thinking that profoundly changes our approach to resources and production. Zero Waste is not about recycling and diversion from landfills but about restructuring production and distribution systems to prevent waste from being manufactured in the first place. The materials that are still required in these re-designed, resource-efficient systems will be reused many times as the products that incorporate them are reused. Deconstruction can be described as construction in reverse. It involves carefully taking apart a building to maximize the reuse of materials, thereby reducing waste and conserving resources. Deconstruction can capture materials and some components from the millions of buildings that are existing and that were poorly designed for high level reuse but it is not a favored approach from a Zero Waste point of view. Zero Waste favors the design of buildings as assemblages of high level components, not their creation from rough materials such as lumber, cement or plaster. The details are not worked out yet but to the extent that entire rooms, entire walls, roofs or floors or entire utility systems can be pre-built and installed as completed components, that will be the goal of Zero Waste design. Until buildings are built as components capable of later dismantling, deconstruction is a stop-gap process that the United States can use to minimize the waste of building materials. For now, the largest parts that we are able to save tend to be architectural elements, windows, doors, and metals, many of which are being saved and resold by reuse yards such as Urban Ore in Berkeley, California. The main parts that still need to be crushed are wood flooring, brick walls, and structural timbers. The demolition of traditional buildings has been long done by wrecking ball or bulldozer. Social and political artifacts, such as demolition contractor licenses and required permits that can only be satisfied by destruction and discard (with partial recycling of rubble and steel), render the destruction and disposal costs cheaper than deconstruction. Approximately seventy pounds of the waste is generated for about every square foot of the residential building demolition. It is arguable that this is artificial economics, based on the cultural preference for wastefulness and that Zero Waste designs of dismantlable components will ultimately be the cheapest as well as the most conservative way to reuse buildings.

Brick, wood and stone are among the oldest truly recyclable materials used in construction. A historic review of old buildings, barns and bridges clearly shows that brick, stones and timber are reused from older buildings. Some of the oldest structures on the planet are built with materials that were recycled from previous structures. One recent example is the Mayflower Barn at Jordans just north of High Wycombe, UK. The barn is clearly built of reused timbers, possibly sourced from the salvage of the Mayflower ship. It is simply a fact of life that historically materials that could be reused were reused.

In more recent construction, structural timber components, including large timbers, glued laminated beams, floor joists, studs and flooring are some of the most valuable structural components salvaged when a structure is demolished if there is an interest in salvaging. If you need proof, go down to any local construction salvage yard and look at the value of trusses, wood beams, floor joists, studs and flooring. Today they have value when someone saves them.

One of the barriers of reusing structural materials is the bias of building code officials and building departments that discriminate against reusing materials. Codes and building departments require compliance to codes, including the source of materials. Your average contractor cannot just use 100-year-old 2x8 (50x200) salvaged floor joists because the building department requires a graded joist. The contractor then has to find an engineer or wood technologist to verify the material suitability for its use. While the codes technically allow this under alternative method, modern attention to cost usually prevents that option.

Market-based Campaigns

Market-based, legislation-mediated campaigns like Extended Producer Responsibility (EPR) and the Precautionary Principle are among numerous campaigns that have a Zero Waste slogan hung on them by means of claims they will ineluctably lead to policies of Zero Waste. At the moment, there is no evidence that EPR will increase reuse, rather than merely moving discard and disposal into private-sector dumping contracts. The Precautionary Principle is put forward to shift liability for proving new chemicals are safe from the public (acting as guinea pig) to the company introducing them. As such, its relation to Zero Waste is dubious. Likewise, many organizations, cities and counties have embraced a Zero Waste slogan while pressing for none of the key Zero Waste changes. In fact, it is common for many such to simply state that recycling is their entire goal. Many commercial or industrial companies claim to embrace Zero Waste but usually mean no more than a major materials recycling effort, having no bearing on product redesign. Examples include Staples, Home Depot, Toyota, General Motors and computer take-back campaigns. Earlier social justice campaigns have successfully pressured McDonald's to change their meat purchasing practices and Nike to change its labor practices in Southeast Asia. Those were both based on the idea that organized consumers can be active participants in the economy and not just passive subjects. However, the announced and enforced goal of the public campaign is critical. A goal to reduce waste generation or dumping through greater recycling will not achieve a goal of product redesign and so cannot reasonably be called a Zero Waste campaign.

How to Achieve

National and provincial governments often set targets and may provide some funding, but on a practical level, waste management programs (e.g. pickup, dropoff, or containers for recycling and composting) are usually implemented by local governments, possibly with regionally shared facilities.

Reaching the goal of zero waste requires the products of manufacturers and industrial designers to be easily disassembled for recycling and incorporated back into nature or the industrial system; durability and repairability also reduce unnecessary churn in the product lifecycle. Minimizes packaging also solves many problems early in the supply chain. If not mandated by government, choices by retailers and consumers in favor of zero-waste-friendly products can influence production. More and more schools are motivating their students to live a different live and rethink every polluting step they may take. To prevent material from becoming waste, consumers, businesses, and non-profits must be educated in how to reduce waste and recycle successfully. In the recent years, the zero waste movement has experienced a significant increase in followers, thanks to eco-activists such as Bea Johnson, author of "Zero Waste Home", and Lauren Singer, an eco activist living in New York and running social media channels to spread the awareness about the movement.

Zero Waste Hierarchy

The "Zero Waste Hierarchy" describes a progression of policies and strategies to support the Zero Waste system, from highest and best to lowest use of materials. It is designed to be applicable to all audiences, from policy-makers to industry and the individual. It aims to provide more depth to the internationally recognized 3Rs (Reduce, Reuse, Recycle); to encourage policy, activity and investment at the top of the hierarchy; and to provide a guide for those who wish to develop systems or products that move us closer to Zero Waste. It enhances the Zero Waste definition by providing guidance for planning and a way to evaluate proposed solutions. All over the world, in some form or another, a pollution prevention hierarchy is incorporated into recycling regulations, solid waste management plans, and resource conservation programs. In Canada, a pollution prevention hierarchy otherwise referred to as the Environmental Protection Hierarchy was adopted. This Hierarchy has been incorporated into all recycling regulations within Canada and is embedded within all resource conservation methods which all government mandated waste prevention programs follow. While the intention to incorporate the 4th R (recovery) prior to disposal was good, many organizations focused on this 4th R instead of the top of the hierarchy resulting in costly systems designed to destroy materials instead of systems designed to reduce environmental impact and waste. Because of this, along with other resource destruction systems that have been emerging over the past few decades, Zero Waste Canada along with the Zero Waste International Alliance have adopted the only internationally peer reviewed Zero Waste Hierarchy that focuses on the first 3Rs; Reduce, Reuse and Recycle including Compost.

Zero Waste Jurisdictions

Various governments have declared zero waste as a goal, including:

- Brazil:
  - Florianópolis, Santa Catarina.
- Canada:
  - Vancouver.
- Italy:
  - Capannori, Tuscany.
- Japan:
  - Kamikatsu, Tokushima.
- United States:
  - Austin, Texas.
  - Boulder, Colorado.
  - Fort Collins, Colorado.
  - San Francisco, California.

An example of network governance approach can be seen in the UK under New Labour who proposed the establishment of regional groupings that brought together the key stakeholders in waste management (local authority representatives, waste industry, government offices etc.) on a voluntary basis. There is a lack of clear government policy on how to meet the targets for diversion from landfill which increases the scope at the regional and local level for governance networks. The overall goal is set by government but the route for how to achieve it is left open, so stakeholders can coordinate and decide how best to reach it.

Zero Waste is a strategy promoted by environmental NGOs but the waste industry is more in favour of the capital intensive option of energy from waste incineration. Research often highlights public support as the first requirement for success. In Taiwan, public opinion was essential in changing the attitude of business, who must transform their material use pattern to become more sustainable for Zero Waste to work.

The latest development in Zero Waste is the city of Masdar in Abu Dhabi which promises to be a Zero Waste city. Innovation and technology is encouraged by government creating an innovation friendly environment without being prescriptive. To be a successful model of sustainable urban development it will also require the involvement and co-operation from all members of society emphasizing the importance of network governance.

RESOURCE EFFICIENCY

Resource efficiency is the maximising of the supply of money, materials, staff, and other assets that can be drawn on by a person or organization in order to function effectively, with minimum wasted (natural) resource expenses. It means using the Earth's limited resources in a sustainable manner while minimising environmental impact.

Motivation

A 2014 report by The Carbon Trust suggested that resource challenges are intensifying rapidly – for example, there could be a 40% gap between available water supplies and water needs by 2030, and some critical materials could be in short supply as soon as 2016. These challenges could lead to disruptions to supply, growing regulatory requirements, volatile fluctuation of prices, and may ultimately threaten the viability of existing business models.

Related Concepts

Resource efficiency measures, methods, and aims are quite similar to those of resource productivity/resource intensity and of the slightly more environment-inclined concept of ecological efficiency/eco-efficiency.

Possible Approaches

To achieve and optimize natural resource and energy efficiency, several sustainable economical or production schemes have been proposed over the course of the last 50 years: circular economy, cradle-to-cradle- or regenerative design, as well as biomimetics principles, just to name a few. Common to all of them is built-in sustainability, in which (non-renewable) resource-wasting is ruled out by design. They are generally built to be holistic, robustly self-sustaining and respecting the carrying capacity of the economic or ecological system.

Resource use Measurement and Identification of Hotspots

A key tool in resource efficiency is measuring different aspects of resource use (e.g. carbon footprint, water footprint, land footprint or material use), then identifying 'hot spots' where the most resources are used or where there are the best opportunities to reduce this resource use. For example, WRAP has published information on hotspots for 50 grocery products likely to contribute most to the environmental impacts associated with UK household consumption. WRAP have created a range of tools and guides to help improve business resource efficiency.

POLLUTER PAYS PRINCIPLE

In environmental law, the polluter pays principle is enacted to make the party responsible for producing pollution responsible for paying for the damage done to the natural environment. It is regarded as a regional custom because of the strong support it has received in most Organisation for Economic Co-operation and Development (OECD) and European Union countries. It is a fundamental principle in US environmental law.

Applications in Environmental Law

The polluter pays principle underpins environmental policy such as an ecotax, which, if enacted by government, deters and essentially reduces greenhouse gas emissions. This principle is based on the fact that as much as pollution is unavoidable, the person or industry that is responsible for the pollution must pay some money for the rehabilitation of the polluted environment.

Australia

The state of New South Wales in Australia has included the polluter pay principle with the other principles of ecologically sustainable development in the objectives of the Environment Protection Authority.

European Union

The polluter pays principle is set out in the Treaty on the Functioning of the European Union and Directive 2004/35/EC of the European Parliament and of the Council of 21 April 2004 on environmental liability with regard to the prevention and remedying of environmental damage is based on this principle. The directive entered into force on 30 April 2004; member states were allowed three years to transpose the directive into their domestic law and by July 2010 all member states had completed this.

France

In France, the Charter for the Environment contains a formulation of the polluter pays principle:

> Everyone shall be required, in the conditions provided for by law, to contribute to the making good of any damage he or she may have caused to the environment.

Ghana

In Ghana, the polluter pays principle was adopted in 2011.

Sweden

The polluter pays principle is also known as extended producer responsibility (EPR). This is a concept that was probably first described by Thomas Lindhqvist for the Swedish government in 1990. EPR seeks to shift the responsibility of dealing with waste from governments (and thus, taxpayers and society at large) to the entities producing it. In effect, it internalised the cost of waste disposal into the cost of the product, theoretically meaning that the producers will improve the waste profile of their products, thus decreasing waste and increasing possibilities for reuse and recycling.

The Organisation for Economic Co-operation and Development defines extended producer responsibility as:

> A concept where manufacturers and importers of products should bear a significant degree of responsibility for the environmental impacts of their products throughout the product life-cycle, including upstream impacts inherent in the selection of materials for the products, impacts from manufacturers' production process itself, and downstream impacts from the use and disposal of the products. Producers accept their responsibility when designing their products to minimise life-cycle environmental impacts, and when accepting legal, physical or socio-economic responsibility for environmental impacts that cannot be eliminated by design.

Switzerland

The waste management in Switzerland is based on the polluter pays principle. Bin bags (for municipal solid waste) are taxed with pay-per-bag fees in three quarters of the communes (and the recycling rate doubled in twenty years).

United Kingdom

The Environmental Damage (Prevention and Remediation) Regulations 2009 (for England) and the Environmental Damage (Prevention and Remediation) (Wales) Regulations 2009 (for Wales) established the operation of the polluter pays principle.

United States

The principle is employed in all of the major US pollution control laws: Clean Air Act, Clean Water Act, Resource Conservation and Recovery Act (solid waste and hazardous waste management), and Superfund (cleanup of abandoned waste sites).

Some eco-taxes underpinned by the polluter pays principle include:

- The Gas Guzzler Tax for motor vehicles.
- Corporate Average Fuel Economy (CAFE), a "polluter pays" fine.
- The Superfund law requires polluters to pay for cleanup of hazardous waste sites, when the polluters can be identified.

In 2003, an amendment was passed negating the "polluter pays" provision of the Florida Constitution.

Limitations of Polluter Pays Principle

The US Environmental Protection Agency (EPA) has observed that the polluter pays principle has typically not been fully implemented in US laws and programs. For example, drinking water and sewage treatment services are subsidized and there are limited mechanisms in place to fully assess polluters for treatment costs.

Zimbabwe

The Zimbabwe Environmental Management Act of 2002 prohibits the discharge of pollutants into the environment. In line with the "Polluter Pays" principle, the Act requires a polluter to meet the cost of decontaminating the polluted environment.

EXTENDED PRODUCER RESPONSIBILITY

Tires are an example of products subject to extended producer responsibility in many industrialized countries.

In the field of waste management, extended producer responsibility (EPR) is a strategy to add all of the environmental costs associated with a product throughout the product life cycle to the market price of that product. Extended producer responsibility legislation is a driving force behind the adoption of remanufacturing initiatives because it "focuses on the end-of-use treatment of consumer products and has the primary aim to increase the amount and degree of product recovery and to minimize the environmental impact of waste materials".

The concept was first formally introduced in Sweden by Thomas Lindhqvist in a 1990 report to the Swedish Ministry of the Environment. In subsequent reports prepared for the Ministry, the following definition emerged: "EPR is an environmental protection strategy to reach an environmental objective of a decreased total environmental impact of a product, by making the manufacturer of the product responsible for the entire life-cycle of the product and especially for the take-back, recycling and final disposal.

EPR uses financial incentives to encourage manufacturers to design environmentally friendly products by holding producers responsible for the costs of managing their products at end of life. This policy approach differs from product stewardship, which shares responsibility across the chain of custody of a product, in that it attempts to relieve local governments of the costs of managing certain priority products by requiring manufacturers to internalize the cost of recycling within the product price. EPR is based on the principle that manufacturers (usually brand owners) have the greatest control over product design and marketing and have the greatest ability and responsibility to reduce toxicity and waste.

EPR may take the form of a reuse, buyback, or recycling program. The producer may also choose to delegate this responsibility to a third party, a so-called producer responsibility organization (PRO), which is paid by the producer for used-product management. In this way, EPR shifts the responsibility for waste management from government to private industry, obliging producers, importers and sellers to internalise waste management costs in their product prices and ensure the safe handling of their products.

A good example of a producer responsibility organization is PRO Europe S.P.R.L. (Packaging Recovery Organisation Europe), founded in 1995, the umbrella organization for European packaging and packaging waste recovery and recycling schemes. Product stewardship organizations like PRO Europe are intended to relieve industrial companies and commercial enterprises of their individual obligation to take back used products through the operation of an organization which fulfills these obligations on a nationwide basis on behalf of their member companies. The aim is to ensure the recovery and recycling of packaging waste in the most economically efficient and ecologically sound manner. In many countries, this is done through the Green Dot trademark of which PRO Europe is the general licensor. In twenty-five nations, companies are now using the Green Dot as the financing symbol for the organization of recovery, sorting and recycling of sales packaging.

Take-back

In response to the growing problem of excessive waste, several countries adopted waste management policies in which manufacturers are responsible for taking back their products from end users at the end of the products' useful life, or partially financing a collection and recycling infrastructure. These policies were adopted due to the lack of collection infrastructure for certain products that contain hazardous materials, or due to the high costs to local governments of providing such collection services. The primary goals of these take-back laws therefore are to partner with the private sector to ensure that all waste is managed in a way that protects public health and the environment. The goals of take-back laws are to:

- Encourage companies to design products for reuse, recyclability, and materials reduction;
- Correct market signals to the consumer by incorporating waste management costs into product price;
- Promote innovation in recycling technology.

Take-back programs help promote these goals by creating incentives for companies to design products that minimize waste management costs, to design products that contain safer materials (so they do not need to be managed separately), or to design products that are easier to recycle and reuse (so recycling becomes more profitable). The earliest take-back activity began in Europe, where government-sponsored take-back initiatives arose from concerns about scarce landfill space and potentially hazardous substances in component parts. The European Union adopted a directive on Waste Electrical and Electronic Equipment (WEEE). The purpose of this directive is to prevent the production of waste electronics and also to encourage reuse and recycling of such waste. The directive requires the Member States to encourage design and production methods that take into account the future dismantling and recovery of their products. These take-back programs have been adopted in nearly every OECD country. In the United States, most of these policies have been implemented at the state level.

Plastic Bags

Recycling, banning, and taxation fails to adequately reduce the pollution caused by plastic bags. An alternative to these policies would be to increase extended producer responsibility. In the US, under the Clinton presidency, the President's Council on Sustainable Development suggested EPR in order to target different participants in the cycle of a product's life. This can, however, make the product more expensive since the cost must be taken into consideration before being put on the market, which is why it is not widely used in the United States currently. Instead, there is banning or taxation of plastic bags, which puts the responsibility on the consumers. In the United States, EPR has not successfully been made mandatory, instead being voluntary. What has been recommended is a comprehensive program which combines taxation, producer responsibility, and recycling to combat pollution.

Electronics

Many governments and companies have adopted extended producer responsibility to help address the growing problem of e-waste — used electronics contain materials that cannot be safely thrown away with regular household trash. In 2007, according to the Environmental Protection Agency, people threw away 2.5 million tons of cell phones, TVs, computers, and printers. Many governments have partnered with corporations in creating the necessary collection and recycling infrastructure. Some argue that local and manufacturer-supported extended producer responsibility laws give manufacturers greater responsibility for the reuse, recycling, and disposal of their own products.

The kinds of chemicals found in e-waste that are particularly dangerous to human health and the environment are lead, mercury, brominated flame-retardants, and cadmium. Lead is found in the screens of phones, TVs and computer monitors and can damage kidneys, nerves, blood, bones, reproductive organs, and muscles. Mercury is found in flat screen TVs, laptop screens, and fluorescent bulbs, and can cause damage to the kidneys and the nervous system. Brominated flame-retardants found in cables and plastic cases can cause cancer, disruption of liver function, and nerve damage. Cadmium is found in rechargeable batteries and can cause kidney damage and cancer. Poorer countries are dumping grounds for e-waste as many governments accept money for disposing this waste on their lands. This causes increased health risks for people in these countries, especially ones who work or live close to these dumps.

In the United States, 25 states have implemented laws that require the recycling of electronic waste. Of those, 23 have incorporated some form of extended producer responsibility into their laws. According to analysis done by the Product Stewardship Institute, some states have not enacted EPR laws because of a lack of recycling infrastructure and funds for proper e-waste disposal. In contrast, according to a study of EPR legislation done by the Electronics TakeBack Coalition, states that have seen success in their e-waste recycling programs have done so because they have developed a convenient e-waste infrastructure or the state governments have instituted goals for manufacturers to meet.

Advocates for EPR also argue that including "high expectations for performance" into the laws, and ensuring that those are only minimum requirements, contribute to making the laws successful. The larger the scope of products that can be collected, the more e-waste will be disposed of properly.

Similar laws have been passed in other parts of the world as well. The European Union has taken steps to address some electronic waste management issues. They have restricted the use of harmful substances in member countries and have made it illegal to export waste.

The Chinese laws regarding e-waste are similar to the ones in the EU, but they focus on banning the import of e-waste. This has proven to be difficult, however, because illegal smuggling of waste still occurs in the country. In order to dispose of e-waste in China today, a license is required and plants are held responsible for treating pollution.

Advantages

When producers face either the financial or physical burden of recycling their electronics after use, they may be incentivized to design more sustainable, less toxic, and more easily recyclable electronics. Using fewer materials and designing products to last longer can directly reduce producers' end-of-life costs. Thus, extended producer responsibility is often cited as one way to fight planned obsolescence, because it financially encourages manufacturers to design for recycling and make products last longer.

Disadvantages

Some people have concerns about extended producer responsibility programs for complex electronics that can be difficult to safely recycle, such as lithium-ion polymer batteries. Others worry that such laws could increase the cost of electronics because producers would add recycling costs into the initial price tag. When companies are required to transport their products to a recycling facility, it can be expensive if the product contains hazardous materials and does not have a scrap value, such as with CRT televisions, which can contain up to five pounds of lead. Organizations and researchers against EPR claim that the mandate would slow innovation and impede technological progress.

Other critics are concerned that manufacturers may use takeback programs to take secondhand electronics off the reuse market, by shredding rather than reusing or repairing goods that come in for recycling. Another argument against EPR is that EPR policies are not accelerating environmentally-friendly designs because "manufacturers are already starting to moving toward reduced material-use per unit of output, reduced energy use in making and delivering each product, and improved environmental performance."

The Reason Foundation argues that EPR is not clear in the way fees are established for the particular recycling processes. Fees are set in place to help incentivize recycling, but this may deter the use of manufacturing with better materials for the different electronic products. There are not set fees for certain materials, so confusion occurs when companies do not know what design features to include in their devices.

Implementation

EPR has been implemented in many forms, which may be classified into three major approaches:

- Mandatory.
- Negotiated.
- Voluntary.

It is perhaps because of the tendency of economic policy in market-driven economies not to interfere with consumers' preferences that the producer-centric representation is the dominant form of viewing the environmental impacts of industrial production: in statistics on energy, emissions, water, etc., impacts are almost always presented as attributes of industries ("on-site" or "direct" allocation) rather than as attributes of the supply chains of products for consumers. On a smaller scale, most existing schemes for corporate sustainability reporting include only impacts that arise out of operations controlled by the reporting company, and not supply-chain impacts According to this world view, "upstream and downstream environmental impacts are allocated to their immediate producers. The institutional setting and the different actors' spheres of influence are not reflected".

On the other hand, a number of studies have highlighted that final consumption and affluence, especially in the industrialised world, are the main drivers for the level and growth of environmental pressure. Even though these studies provide a clear incentive for complementing producer-focused environmental policy with some consideration for consumption-related aspects, demand-side measures to environmental problems are rarely exploited.

The nexus created by the different views on impacts caused by industrial production is exemplified by several contributions to the discussion about producer or consumer responsibility for greenhouse gas emissions. Emissions data are reported to the IPCC as contributions of producing industries located in a particular country rather than as embodiments in products consumed by a particular population, irrespective of productive origin. However, especially for open economies, taking into account the greenhouse gases embodied in internationally traded commodities can have a considerable influence on national greenhouse gas balance sheets. Assuming consumer responsibility, exports have to be subtracted from, and imports added to national greenhouse gas inventories. In Denmark, for example, Munksgaard and Pedersen (2001) report that a significant amount of power and other energy-intensive commodities are traded across Danish borders, and that between 1966 and 1994 the Danish foreign trade balance in terms of CO2 developed from a 7 Mt deficit to a 7 Mt surplus, compared to total emissions of approximately 60 Mt. In particular, electricity traded between Norway, Sweden and Denmark is subject to large annual fluctuations due to varying rainfall in Norway and Sweden. In wet years Denmark imports hydro-electricity whereas electricity from coal-fired power plants is exported in dry years. The official Danish emissions inventory includes a correction for electricity trade and thus applies the consumer responsibility principle.

Similarly, at the company level, "when adopting the concept of eco-efficiency and the scope of an environmental management system stated in for example ISO 14001, it is insufficient to merely report on the carbon dioxide emissions limited to the judicial borders of the company". "Companies must recognise their wider responsibility and manage the entire life-cycle of their products.

Insisting on high environmental standards from suppliers and ensuring that raw materials are extracted or produced in an environmentally conscious way provides a start." A life-cycle perspective is also taken in EPR frameworks: "Producers of products should bear a significant degree of responsibility (physical and financial) not only for the environmental impacts of their products downstream from the treatment and disposal of their product, but also for their upstream activities inherent in the selection of materials and in the design of products." "The major impetus for EPR came from northern European countries in the late 1980s and early 1990s, as they were facing severe landfill shortages. EPR is generally applied to post-consumer wastes which place increasing physical and financial demands on municipal waste management."

EPR has rarely been consistently quantified. Moreover, applying conventional life cycle assessment, and assigning environmental impacts to producers and consumers can lead to double-counting. Using input-output analysis, researchers have attempted for decades to account for both producers and consumers in an economy in a consistent way. Gallego and Lenzen demonstrate and discuss a method of consistently delineating producers' supply chains, into mutually exclusive and collectively exhaustive responsibilities to be shared by all agents in an economy. Their method is an approach to allocating responsibility across agents in a fully inter-connected circular system. Upstream and downstream environmental impacts are shared between all agents of a supply chain – producers and consumers.

Examples:

Auto Recycling Nederland (ARN) is a producer responsibility organisation (PRO) that organises vehicle recycling in the Netherlands. An advanced recycling fee is charged to those who purchase a new vehicle and is used to fund the recycling of it at the end of its useful life. The PRO was set up to satisfy the European Union's End of Life Vehicles Directive.

The Swiss Association for Information, Communication and Organisational Technology (SWICO), an ICT industry organisation, became a PRO to address the problem of electronic waste.

The Canada-Wide Action Plan for Extended Producer Responsibility (CAP-EPR) was adopted in Canada in 2009 under the guidance of the Canadian Council of Ministers of the Environment. The CAP-EPR followed years of waste and recycling efforts in Canada that remained largely ineffective as the diversion rates from landfills and incineration persisted. Despite three decades worth of recycling efforts, Canada fell short of many other G8 and OECD countries. Since the CAP-EPR's 2009 inception, most provinces have enforced legislation or restrictions on a wider range of products and materials under EPR programs. "Nine out of ten provinces have since implemented EPR programs or put requirements in place. As a result of these new programs or requirements and expansion of existing ones, almost half of the product categories for Phase 1 are now covered by legislated EPR programs or requirements across Canada."

Results:

In Germany, since the adoption of EPR, "between 1991 and 1998, the per capita consumption of packaging was reduced from 94.7 kg to 82 kg, resulting in a reduction of 13.4%". Furthermore, due to Germany's influence in EPR, the "European Commission developed one waste directive" for all of member states. One major goal was to have all member states recycle "25% of all packaging material" and have accomplished the goal.

In the United States, EPR is gaining popularity "with 40 such laws enacted since 2008. In 2010 alone,

38 such EPR bills were introduced in state legislatures across the United States, and 12 were signed into law." However, these laws are only at the state level as there are no federal laws for EPR. So far, "only a handful of states have imposed five to six EPR laws as well as 32 states having at least one EPR law".

PAY AS YOU THROW

Pay as you throw (PAYT) (also called trash metering, unit pricing, variable rate pricing, or user-pay) is a usage-pricing model for disposing of municipal solid waste. Users are charged a rate based on how much waste they present for collection to the municipality or local authority.

A variety of models exist depending on the region and municipality. Waste is measured by weight or size while units are identified using different types of bags, tags, containers or even RFID. Services for waste diversion, like recycling and composting, are often provided free of charge where pay-as-you-throw systems are implemented.

There are three main types of pay-as-you-throw programmes: Full-unit pricing: users pay for all the garbage they want collected in advance by purchasing a tag, custom bag, or selected size container. Partial-unit pricing: the local authority or municipality decides on a maximum number of bags or containers of garbage, with collection paid for by taxes. Additional bags or containers are available for purchase should the user exceed the permitted amount. Variable-rate pricing: users can choose to rent a container of varying sizes (some programs offer up to five), with the price corresponding to the amount of waste generated.

Rationale

The two most traditional approaches to disposing of municipal solid waste are a flat-rate system or municipal taxes. All users pay the same municipal taxes regardless of how much waste they present for pickup. Under the flat-rate system there is no incentive to reduce waste produced.

PAYT is based on two guiding principles of environmental policy: the polluter pays principle and the shared responsibility concept. The rationale for PAYT can be divided into three broad categories:

Economic

Under a PAYT scheme, some or all of the costs of waste management can be removed from property tax bills, providing more independence in the management and financial of residential waste system. Waste management services are then treated just like other utilities such as electricity or water that are charged by unit of consumption.

Environmental

PAYT program are an effective tool in increasing waste separation and recycling, and also encourage waste minimization. The result is significant energy savings from transportation, increases in material recovery from recycling, and reduction in pollution from landfills and incinerators. And thus reduce the load of landfills. PAYT programs also indirectly encourage producers to develop more efficient designs and environmentally friendly product life cycles.

Social

Waste collections costs are distributed more fairly among the population, and in proportion to the amount of waste each user generates. Free riders are no longer able to have their behavior subsidized, and PAYT is said to promote community sustainability. Lower-income families tend to produce less waste and thus pay lower waste collection fees.

Diversion Effects and Risks

Charging for waste can also sometimes result in illegal dumping (fly-tipping) or the waste being passed to unlicensed or illegal disposal methods. In practice, increased litter, open burning, and illegal dumping have not been observed by most pay-as-you-throw communities, especially when combined with easier recycling and composting. Municipalities in Massachusetts have typically observed a reduction in trash larger than the increase in recycling, but this is believed to be due to people formerly bringing in trash from other communities (where presumably there is no free pickup) to the homes of friends and relatives.

Political Opposition

When there is a change to any established municipal service, public resistance is common. The Massachusetts Department of Environmental Protection recommends making PAYT programs revenue-neutral by reducing taxes at the same time as introducing trash fees so residents don't see the latter as a new cost.

ENERGY RECOVERY FROM WASTE

Energy can be recovered from the organic fraction of waste (biodegradable as well as non-biodegradable) through thermal, thermo-chemical and biochemical methods.

Thermal Conversion	Biochemical Conversion	Thermochemical Conversion	Electrochemical Conversion
• Technology options – **Incineration** • Thermal degradation of waste at high temperatures. • Complete oxidation of waste matter under high temperature and aerated/oxygenated conditions • End products include steam, purified flue gas and ash.	• Technology options – **Biomethanation (Anaerobic Digestion, Fermentation** • Enzymatic decomposition of organic matter by microbial action to produce methane gas, and alcohol etc • Suitable for wastes having high percentage of organic, bio-degradable (putrescible) matter and high level of moisture/ water content	• Technology options – **Pyrolysis or Gasification** • High temperature-driven decomposition of organic matter to produce either heat or fuel oil or gas • Suitable for wastes containing high percentage of organic non-biodegradable matter and low moisture content. • End products can be used for heat applications or further processed chemically, to produce a range of end products.	• Electro decomposition of organics in the presence of microbes as a charge conducting medium. **Refers to Microbial Fuel Cells (MFCs)** • Exploiting the oxidation-reduction machinery of immobilized microbial cells for generating electricity and bio-hydrogen • Technology is at nascent level in India and world-wide.

Anaerobic Digestion or Biomethanation

In this process, the organic fraction of the waste is segregated and fed into a closed container (biogas digester). In the digester, the segregated waste undergoes biodegradation in presence of methanogenic bacteria and under anaerobic conditions, producing methane-rich biogas and effluent. The biogas can be used either for cooking/heating applications, or for generating motive power or electricity through dual-fuel or gas engines, low-pressure gas turbines, or steam turbines. The sludge from anaerobic digestion, after stabilization, can be used as a soil conditioner. It can even be sold as manure depending upon its composition, which is determined mainly by the composition of the input waste.

Combustion Incineration

In this process, wastes are directly burned in presence of excess air (oxygen) at high temperatures (about 800 °C), liberating heat energy, inert gases, and ash. Combustion results in transfer of 65%–80% of heat content of the organic matter to hot air, steam, and hot water. The steam generated, in turn, can be used in steam turbines to generate power.

Pyrolysis or Gasification

Pyrolysis is a process of chemical decomposition of organic matter brought about by heat. In this process, the organic material is heated in absence of air until the molecules thermally break down to become a gas comprising smaller molecules (known collectively as syngas).

Gasification can also take place as a result of partial combustion of organic matter in presence of a restricted quantity of oxygen or air. The gas so produced is known as producer gas. The gases produced by pyrolysis mainly comprise carbon monoxide (25%), hydrogen and hydrocarbons (15%), and carbon dioxide and nitrogen (60%). The next step is to 'clean' the syngas or producer gas. Thereafter, the gas is burned in internal combustion (IC) engine generator sets or turbines to produce electricity.

Landfill Gas Recovery

The waste dumped in a landfill becomes subjected, over a period of time, to anaerobic conditions. As a result, its organic fraction slowly volatilizes and decomposes, leading to production of 'landfill gas', which contains a high percentage of methane (about 50%). It can be used as a source of energy either for direct heating/cooking applications or to generate power through IC engines or turbines.

References

- "Vancouver votes to ban single-use straws, foam cups and take-out containers". CTV News. May 17, 2018. Retrieved July 28, 2018
- Basic-principles, waste, technical-module, a-sustainable-design-toolkit, building-futures, microsites: hertfordshire.gov.uk, Retrieved 12 August, 2019
- Cox, Jayne; Giorgi, Sarah; Veronica, Sharp; Strange, Kit; Wilson, David C.; Blakey, Nick (2010). "House-hold waste prevention - a review of evidence." Waste Management & Research. 28 (3): 193 219. Doi:10.1177 /0734242x10361506
- Process, pro, wte, ae, ref: eai.in, Retrieved 13 January, 2019
- "Opportunities in a resource constrained world: How business is rising to the challenge". The Carbon Trust. February 2014. Retrieved 22 July 2014

4

Waste Handling and Disposal

Some of the different processes which are a part of waste handling and disposal are waste sorting, kerbside collection, incineration and waste compaction. This chapter closely examines these methods of waste handling and disposal to provide an extensive understanding of the subject.

When we talk about solid waste management, there are many aspects of how such waste needs to be collected, treated, disposed of and other aspects. Today all modern towns and cities are facing the challenge of handling growing amounts of solid wastes and efficient management of such wastes including handling and transportation which is a subject of elaborate planning and management of the local authorities.

Waste collection varies from country to country, region to region. The collection methods depends on the expanse of areas covered, the landfill areas the garbage needs to be transported to amongst other factors. Usually the local government arranges for the waste handling and transportation in most countries though in some places it is handled by private players. Again, in many rural areas of developing countries there is no formal waste collection method.

In some advanced countries including those in Europe a vacuum based system is used to collect the refuse through underground conduits known as Envac. Here the refuse is transported to a designated area through a vacuum pump system, eliminating the need for manual handling of garbage. Metro Taifun is another vacuum based solution of single or ring line systems.

In many cities such as in Canada wastes are collected from the curbside in neighborhoods. Here the garbage bins are designated as wastes or recyclable materials and people are encouraged to segregate their waste before they drop them into the bins. These are then collected based on a daily schedule. In rural areas people need to transfer their garbage to a transfer station from where it is transported to a landfill area.

Many Asian countries face problem of huge volumes of solid waste management. In many such countries such as Taipei, the government charges taxes as per the volume of garbage produced and issue garbage bags in which the waste needs to be collected and deposited by the city people. Such initiatives have helped to reduce the waste accumulation and encouraged people to recycle and reuse.

The ArrowBio system is commonly used in many countries such as Australia, Greece, Mexico, UK or California. Here the trash is collected in trucks and by mechanical methods such as gravitational screening, settling and hydro mechanical shredding. Huge volumes of wastes are segregated, salvaged for recyclables and then turned into biogas or compost.

Thus, different countries and cities have different means of collecting wastes and transporting it but it still remains a huge responsibility on the civic authorities and they need the active help of the public in the growing waste management issues.

WASTE SORTING

Waste sorting is a simple process that requires neither a lot of time, nor any special resources. While sorting waste, the key is to separate from household waste the types of waste that can be recycled.

"Zala josta" suggests dividing the total waste stream into the following groups:

- Packaging materials (plastic, glass, paper/cardboard);
- Metal;
- Batteries;
- Electrical equipment;
- Environmentally hazardous waste;
- Medical waste;
- Biodegradable waste;
- Reusable things;
- Bulky waste and construction waste;
- Household waste not suitable for recycling.

Plastic Packaging Waste

These are mostly empty PET bottles, containers, boxes, film and bags. How to collect, prepare for recycling by washing and removing the impurities, such as food residues, flattening as much as possible. Suitable for recycling, flattened plastic beverage bottles, plastic bags, film, plastic boxes. Not suitable for recycling, dirty plastic products, oil bottles, yogurt, sour cream or margarine containers, toys, household goods, household items. Deliver any separate collection container for plastic waste (usually yellow) or to a specially arranged ECO site.

Glass Packaging Waste

They are all kinds of glass jars, containers, bottles. How to collect, prepare for recycling by washing and removing the impurities, such as food residues. Suitable for recycling, glass jars, bottles, window glass. Not suitable for recycling, mirror glass, auto glass, oil bottles, balsam bottles (of clay). Deliver all the jars and bottles to any sortes collection container for glass waste (usually green) or a specially arranged ECO site. Window glass should be usually delivered to EKO sites.

Paper and Cardboard Wastes

They are all paper packaging, cardboard, newspapers, magazines, books, note paper, egg boxes, and other paper/cardboard products. How to collect, prepare for recycling by removing the impurities and flattening as much as possible. Suitable for recycling, newsprint, cardboard, writing paper, books, unsorted paper, all kinds of paper/cardboard products, also stapled. Not suitable for recycling, paper and cardboard that is wet, dirty, containing food impurities are laminated. Deliver any separate collection container for paper waste (usually blue) or a specially arranged ECO site. Paper can also be delivered to paper-mill "Ligatne" by signing an agreement on paper waste removal.

Metal Waste

These are all types of metal products, and metal packaging such as cans or tins. How to collect, prepare for recycling by removing impurities such as food residues, flattening to possible extent. Suitable for recycling, all kinds of metal products free of contaminants. Not suitable for recycling, greasy, dirty metal packages. Deliver to any EKO site or specialized scrap purchasing points in exchange for an appropriate fee.

Waste Batteries and Accumulators

They are all kinds of batteries and accumulators of various electrical devices. How to collect, collected separately from other waste, keeping away from moisture. Suitable for recycling, any batteries and rechargeable batteries, and batteries that have leaked. Deliver, any specialized battery collection container, which is located in the battery POS - stores, as well as educational institutions, government bodies and specialized ECO sites.

Deteriorated Electrical Appliances

They are any (big or small) electrical and electronic devices that are operated by a power source. How to collect, prepare for recycling by disconnecting from the power supply. Before scrapping it shall not be disassembled by oneself. Suitable for recycling, any type of electrical equipment. Deliver: any EKO site. Take the opportunity to get in touch with collection companies of deteriorated electric equipment, which will remove free of charge the deteriorated appliances from your home. If you purchase new electrical equipment, according to the legislative provisions the seller shall free of charge accept from you the same kind of obsolete electrical equipment and ensure its utilization.

Environmentally Hazardous Waste

They are, waste industrial oils, oil filters, used car tires How to collect, collect in specialized containers provided by the manager of environmentally hazardous products on a contract basis. Tires are collected without any special rules. Suitable for recycling, environmentally hazardous waste free of contaminants. Deliver a specially arranged ECO site or a point of collection of environmentally hazardous products. From private customers without charge there are accepted up to 4 car tires. If you want to deliver more than 4 tires, then it is a paid service.

Medical Waste

It is any kind of medication, which has expired. How to collect in a separate bag. Ensure that bottles are firmly sealed. Suitable for disposal, any medication, which has expired or, which is no longer used. Do not rinse medication down the sewer system, or donate unused medication to others. Deliver any point of collection of hazardous waste or a nearby pharmacy.

Biodegradable Waste

They are any waste of natural origin - fruit and vegetable leftovers, egg shells, biopolymers packaging, also wet paper waste. How to collect in a separate container preparing for composting. Not suitable for composting, dairy and meat products leftovers, which during the decomposition will distribute an unpleasant odor and promote bacterial growth. Deliver compost at the back garden, creating a compost pile or compost box.

Reusable Things

They are clothing, shoes, household goods, textiles, furniture and other things that can be reused. Deliver donate to charity, the Red Cross or orphanages, homes for the elderly. Old blankets and worn clothing will serve animal shelters.

Bulky Waste and Construction Waste

They are old and broken-down shell of soft furniture, any redecoration waste. Deliver, you should contact the waste management with a request to provide you with a special container to collect the mentioned waste. This is a paid service.

Household Waste not Suitable for Recycling

Everything else that could not be classified to any of the recyclable material groups is household waste. It is brought to a regular unsorted waste container - it will be stored in a waste site and will not be recycled.

Tips on Sorting Waste

1. Decline without regrets

Always refuse any material you don't need, especially packaging. Reducing the volume of trash is the best way to reduce the amount of solid waste in the landfill.

Some stores already have places for you to leave your packaging at the time of purchase (discard your toothpaste box at the supermarket checkout, for instance).

2. To wash or not to wash

Cleaning the leftover food from the discards will help people who work in the recycling plant. Your yogurt cup will take days or even weeks to reach the recycling site. Have you thought about the smell? To avoid wasting water and to make the task easier, leave the recyclables in the sink so they can be washed with the water you're using for the dishes. Another approach is to take advantage of already-used napkins to clean the items.

3. Compost almost everything damp or squishy

This type of waste accounts for half the volume of all solid waste, fills up landfills and produces methane gas, which is 20 times more polluting than carbon dioxide. But a lot of it can be composted. Fruit and vegetable peels, coffee grounds and even wet paper can be turned into fertilizer.

4. Do you crumple up paper when throwing it in the garbage can?

Avoid folding or crumpling paper before throwing it away. The reason is easy to understand: The more intact paper is, the more money it is worth, as the (cellulose) fibers are stronger, increasing the number of times it can be recycled.

5. Worthless paper

On the other hand, there are many types of paper that cannot be recycled: greaseproof and plasticized paper, receipts from credit card machines or supermarkets (avoid requesting them), paper with adhesive glue (separate the glue from the paper), dirty toilet paper, carbon paper and tax coupons. Used paper napkins and paper towels don't belong in your recycled paper bin, either, but they can be composted. So can greasy pizza boxes; that way they won't dirty the rest of the recyclable paper.

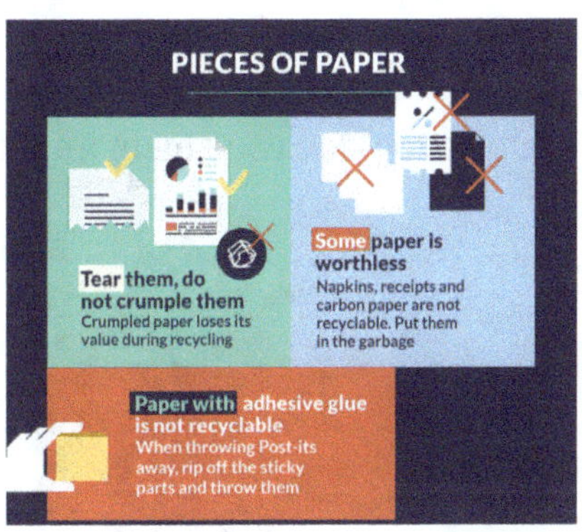

6. Costly Glass

For the future, to avoid the cost of recycling glass, we must fight to increase the use of returnable containers. Until then, reuse glass to the maximum. Whenever possible, avoid long neck bottles.

They are not returnable and the recycling process (which shatters them into shards) is still complex and expensive.

7. Keep tabs on those tabs

Do not remove the tabs from your soda or beer cans. They can get lost amidst the refuse and end up not being recycled.

8. Squeeze that toothpaste

The tube is made of low-density polyethylene, which is a type of plastic. Therefore, try to squeeze as much toothpaste from it as possible and place it in the recycling bin with the other plastic items.

9. Snack packaging

You may not know this, but these products often have in their packaging a type of plastic which is covered by a thin layer of aluminum. It can be recycled. The problem is that the process is expensive, and only a few companies are able to pay for it. Find out how it works in your area. And, try to avoid buying products with this type of packaging.

10. Goodbye polystyrene tray

When you buy electronics packed with styrofoam, recycle it. This kind of styrofoam has market value and therefore the process of recycling it is affordable. That polystyrene tray with the plastic wrap used to store fish, meat and even vegetables is another story. Because it has little market value, it is expensive to recycle. So, take containers made of glass or aluminum to your butcher or fishmonger and ask them to use these to pack your meat and fish. Fruits and vegetables do not need packaging.

11. Warning

Hazardous waste cannot be thrown into the regular waste stream. Like the consumers who buy them, companies producing toxic items must take responsibility for their disposal. Thus, they must provide collection sites, where the materials can be safely picked up and recycled. Some examples are:

- Bottles and packs of medicine: The blister packs cannot be recycled normally and should not be placed in the bin with the rest of your plastics. Like some medicine bottles, they contain remnants of chemical compounds that can contaminate the environment. They're also, in some countries, made with a highly toxic plastic that is one of the least recyclable. So, you must take them to collection sites, which can be found in many drugstores. Depending on where you live, some municipalities collect all these items at "medical collection sites," along with unused medicines.
- Batteries, lamps and ink cartridges: There are collection sites at supermarkets and specialized stores for all of these items. In the case of the ink cartridge, it's best to let the ink dry for at least 24 hours – this makes its container easier to recycle.
- Toxic Brew: Coffee capsules are very difficult to recycle because, like the snack packaging mentioned above, they contain a couple different materials: aluminum and plastics. Fortunately, the major brands selling these products offer collection sites at their stores. But consider doing yourself and the planet a favor and buying a coffee machine that uses grounds.

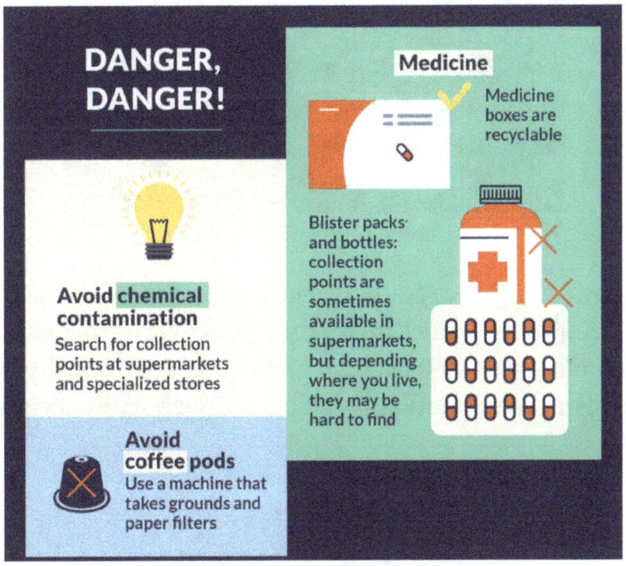

12. Do you know what garbage looks like?

There are objects that are not recyclable, and some of them will surprise you.

Besides all the non-recyclable paper mentioned above, film wrap, paper clips, steel sponges and tape are also considered garbage. They go straight to the landfill, along with the non-compostable damp garbage (dirty diapers, used kitty litter, all that lovely aromatic stuff).

KERBSIDE COLLECTION

Kerbside collection or curbside collection is a service provided to households, typically in urban and suburban areas of collecting and disposing of household waste and recyclables. It is usually accomplished by personnel using purpose built vehicles to pick up household waste in containers acceptable to or prescribed by the municipality.

Kerbside collection.

Waste Management and Resource Recovery

Glass bottles and jars for collection.

Kerbside collection is today often referred to as a strategy of local authorities to collect recyclable items from the consumer. Kerbside collection is considered a low-risk strategy to reduce waste volumes and increase the recycling rates. Recyclable materials are typically collected in large wheelie bins, coloured plastic bags, or small open plastic boxes, specifically designated for content.

Discarded Christmas trees awaiting collection.

Recyclable materials that may be separately collected from municipal waste include:

Biodegradable waste component:

- Green waste,
- Kitchen and food waste.

Recyclable materials, depending on location:

- Office paper,
- Newsprint,
- Paperboard,
- Cardboard,
- Corrugated fiberboard.
- Plastics (number 1 PET, number 2 HDPE natural and colored, number 3 PVC narrow-necked containers, number 4 LDPE, number 5 PP, number 6 Polystyrene (however, not expanded polystyrene, an example of recyclable polystyrene may be a plastic yoghurt pot) number 7 other mixed resin plastics).
- Glass,
- Copper,
- Aluminium,
- Steel and Tinplate,
- Co-mingled recyclables- can be sorted by a clean materials recovery facility.

All accepted paper, glass, plastic, and metal recyclables are collected from one wheeled bin.

Kerbside collection of recyclable resources is aimed to recover purer waste streams with higher market value than by other collection methods. If the household residents incorrectly separate the recyclable materials, or put unsuitable or unsafe items in the recycling bin, the whole vehicle load of recycling may have to be sent to landfill or incineration if it is deemed to be contaminated.

Kerbside collection and household recycling schemes are also being used as tools by many local authorities to increase the public's awareness of their waste production.

Kerbside collection is commonly considered to be completely environmentally friendly. This may not necessarily be true as it leads to an increased number of waste collection vehicles on the road, in themselves contributing to traffic congestion and global warming through exhaust emissions of carbon dioxide until the time of their conversion to clean energy, such as to use electric vehicles that run on batteries.

New and emerging waste treatment technologies such as mechanical biological treatment may offer an alternative to kerbside collection through automated separation of waste in recycling factories.

Usage by Country

Australia

Residential kerbside collection is carried out by local governments, with some exceptions, e.g. some large apartment complexes may have alternate arrangements with commercial providers. Available services and details vary from council to council. Councils generally provide residents with wheelie bins for Kerbside collection of normal waste which is collected weekly or fortnightly. They also may have less frequent kerbside collection of bulkier waste.

Councils provide their residents with two or three wheelie bins, depending on the council, with some councils having different options for different properties. The two bin system consists of a recycling bin (usually 240 litre) for co-mingled recyclables, and a general waste bin which is often smaller (e.g. 140 litre, 120 litre or 80 litre). The three bin system consists of the above two bins plus a green waste bin (usually 240 litre). Not all councils have a green waste bin collection service. Many councils provide the option of larger bins, smaller bins, or additional bins.

A wide variety of hard plastics, glass bottles and jars, steel cans, aluminium cans, paper and cardboard can be placed into the recycling bin. The green waste bin can be used for garden organics (e.g. small branches, leaves, grass clippings), however, many councils allow food scraps, used tissues and other organics to be placed in the green waste bin and more are doing so over time. The council may turn the green waste into mulch (garden waste collection only) or compost and extract energy (food organics and garden organics). Details of what can and cannot be placed into each bin vary by council.

Most councils follow a standard colour scheme for their wheelie bins, specified in Australian standard AS4123. According to the standard, general waste bins have a red lid, recycling bins have a yellow lid, green waste bins have a lime green lid, and all these bins have a dark green or black body. Not all councils follow this colour scheme. For example, recycle bins in some councils have a blue body and yellow lid.

Bins are emptied according to one of several patterns. Generally speaking, general waste bins are emptied weekly while recycling bins and green waste bins are emptied fortnightly on alternate weeks. Many councils who have introduced food scrap recycling have switched to emptying green waste bins weekly and general waste bins and recycling bins fortnightly on alternate weeks. Some councils empty recycling bins weekly, while others do so only during a certain period like the Christmas and summer holiday period, switching to fortnightly at other times.

Recycling bins are provided at no additional cost, while the general waste bin is either at no additional cost or at an annual cost. The green waste bin, where available, is either provided to all residents, or available as an option to residents, either at an additional annual cost, or a one-off cost, or at no additional cost, depending on the council. Some councils limit the availability of green waste bins (e.g. the City of Cockburn limits them to properties over a certain size). Many councils provide the option of larger bins than the standard ones provided (even larger than 240 litres in some cases) or additional bins at additional annual cost. Some provide the option of a smaller general waste bin at a reduced cost.

Councils also have kerbside collection of bulky waste. There may be different kinds of collection, e.g.:

- Large branches.
- E-waste (e.g. TVs, computers) which the council may recycle.
- Hard rubbish (anything else too bulky for the wheelie bin.

For bulky waste, residents are asked to place items directly on the kerbside. There may be other rules, e.g. what can and cannot be collected, limits on the amount of rubbish that will be collected, etc. that vary from council to council. Collections may occur once or a few times a year on specific dates or date ranges, or on demand with a limit to the number of times per year, depending on the council.

Canada

Halifax Regional Municipality (HRM) in Nova Scotia, Canada, with a population of about 375,000, has one of the most complex kerbside collection programmes in North America. Based on the green cart, it requires residents to self-sort refuse and place different types at the kerb on alternating weeks. As shown in the photo at left, week 1 would see the green cart and optional orange bags used for kitchen waste and other organics such as yard waste. Week 2 would permit non-recoverable waste in garbage bags or cans. Blue bags are used for paper, plastic and metal containers. Together with used grocery bags containing newspapers, they may be placed on the kerb either week. In summer, the green cart is emptied weekly due to the prevalence of flies. HRM has achieved a diversion rate of approximately 60 percent by this method.

Calgary, Alberta has adopted "Curbside" Recycling and uses blue bins. The blue cart programme accepts all types of recycables, including plastics 1-7. It is picked up weekly for the cost of $8.00 per month. This programme is mandatory.

In 1981 Resource Integration Systems (RIS) in collaboration with Laidlaw International tested the first blue box recycling system on 1500 homes in Kitchener, Ontario. Due to the success of the project the City of Kitchener put out a contract for public bid in 1984 for a recycling system citywide. Laidlaw won the bid and continued with the popular blue box recycling system. Today hundreds of cities around the world use the blue box system or a similar variation.

Many Canadian municipalities use "green bins" for kerbside recycling. Others, such as Moncton, use wet/dry waste separation and recovery programmes.

New Zealand

Kerbside collection bins in Dunedin, New Zealand. The yellow-liddied wheelie bin is for non-glass recyclables, and the blue bin is for glass. The two bins are collected on alternating weeks. Official council bags are used for general household waste, and are collected weekly.

In New Zealand, kerbside collection of general refuse and recycling, and in some areas organic waste, is the responsibility of the local city or district council, or private contractors. Practices and collection methods vary widely from council to council and company to company. Some examples of collection are:

- Auckland City Council: Two 240-litre wheelie bins are supplied: a red-lidded bin for general refuse, collected weekly, and a blue-lidded bin for recyclables, collected fortnightly.
- Christchurch City Council: Three wheelie bins are supplied: a 140-litre red-lidded bin for general refuse, a 240-litre yellow-lidded bin for recyclables, and an 80-litre green-lidded

bin for organic waste. The organic waste bins are collected weekly, while the recyclables and general refuse bins are collected on alternating weeks.

- Hamilton City Council and Hutt City Council: A 45-litre bin is supplies for recyclables, collected weekly. General refuse is collected weekly using user-pays official council bags.
- Dunedin City Council, Palmerston North City Council and Wellington City Council: Two bins are supplied: a 45-litre or 70-litre bin for glass, and an 80-litre or 240-litre wheelie bin for non-glass recyclables. These two bins are collected on alternating weeks. General refuse is collected weekly using user-pays official council bags.
- Rodney District Council: A 45-litre bin is supplies for recyclables, collected weekly. There is no council collection of general waste, and all general waste collection is carried out by independent companies.
- Taupo District Council: A 45-litre bin is supplies for recyclables, collected weekly. General refuse is collected weekly using user-pays system of orange tags - one orange tag is to be placed on a standard rubbish bag up to 60 litres capacity, or half an orange sticker can be placed on two supermarket bags tied together.
- Upper Hutt City Council: Recycling is to be placed in plastic bags, with paper and cardboard collected in the first week, and plastic, metal and glass in the second week. General refuse is collected weekly using user-pays official council bags.
- Waitakere City Council: A 140-litre wheelie bin is provided for recyclables, collected fortnightly. General refuse is collected weekly using user-pays official council bags.

By 1996 the New Zealand cities of Auckland, Waitakere, North Shore and Lower Hutt had kerbside recycling bins available. In New Plymouth, Wanganui and Upper Hutt recyclable material was collected if placed in suitable bags. By 2007, 73% of New Zealanders had access to kerbside recycling.

Kerbside collection of organic waste is carried out by the Mackenzie District Council and the Timaru District Council. Christchurch City Council is introducing the system to their kerbside collection. Other councils are carrying out trials.

United Kingdom

In the United Kingdom, the Household Waste Recycling Act 2003 requires local authorities to provide every household with a separate collection of at least two types of recyclable materials by 2010. There has been criticism in the difference of schemes used in the country such as the colour of bins, whether they are bins, boxes or bags, and also the fact that the bins, boxes and bags obstruct the roads and pavements and how the additional collection vehicles and waste collection services needed also produce carbon dioxide emissions. Some find the colour differences confusing, and some people want a national scheme. A typical example is to compare two neighbouring councils in Greater Manchester, Bury Council and Salford City Council. Bury uses blue for cans, plastic and glass, green for paper and cardboard and brown for garden waste. Salford uses blue for paper and card, brown for cans plastic and glass and pink for garden waste. Most councils use grey or black for general waste, with a few exceptions such as Liverpool, which uses purple for general waste, a colour used by no other council.

Another controversial issue in the UK is the frequency of the waste collections. To save money, many councils are cutting the frequency of both general waste and recyclables collections. This has led to problems from larger households, and has led to an increase of overflowing bins and fly tipping. For example, previously, Bury Council collected general waste once a week and recyclables fortnightly. This has now changed to fortnightly for general waste and monthly (every 4 weeks) collection of recyclables.

A few councils are using "forced" recycling, by replacing the large, 240 litre general waste bin with a smaller 180 litre or 140 litre bin, and using the old 240 litre one for recyclables. This may be made worse by fortnightly collections of the "small" bin, and strict rules such as "No extra waste will be collected" and "Bin lids must be fully closed". Stockport Council is a notable user of this scheme. Their recycling rates have substantially increased as a result, but there are usually complaints from household residents. Trafford Council also use a similar scheme, but the small grey bin is emptied every week. In addition, the two named councils, and more, collect food waste together with garden waste, by sending out kitchen caddies and compostable bin liners. These prevent food waste (including meat and fish) from going to landfill or incineration, and to increase the council's recycling rate. The food and garden waste is usually collected weekly or fortnightly, and is taken to an in-vessel composting facility or anaerobic digestion plant, where the biodegradable waste is organically recycled into soil fertiliser to be used on local farms.

In North West England, all the glass collected is used within the UK, around half of the plastics and cans are used in the UK; the rest is sent further afield to Europe or China to be made into new products, and paper and paperboard collected is sent to local paper mills to be reprocessed into newspapers, tissues and paper towels, cardboard and office paper. Once again, some of the paper will be sent further afield.

Some councils only use three bins, i.e. general waste, food and garden waste and mixed recyclables. This means that plastics, cans and glass share the same bin with paper and cardboard. Although this is much easier for the residents, there is more sorting required, and the paper quality is sometimes of a low grade due to food waste contamination or shards of glass in the paper, and so this scheme has been criticised.

Also, most councils require residents to remove all caps and lids from plastic bottles, and thoroughly rinse them out to avoid unpleasant smells or liquid contamination. This is because the caps and lids are made from a different type of plastic (PP) from the bottle (PET/HDPE); if the bottles are collapsed and folded over like toothpaste tubes and caps are screwed back in place, the volume of bottles is greatly reduced, so that more bottles can be contained within the recycling bins. In fact many bottlers, especially bottled water companies, have now designed their bottles to be collapsible; though this message has not been effectively disseminated to the consumer. A collapsible bottle takes between 25% and 33% of the space a non-collapsed bottle.

Labels and neck rings can, however, be left on the bottles and they do not need to be removed. This also means that only plastic bottles can be recycled. Many councils are still trying to remind residents that plastic pots, tubs and trays (yoghurts, desserts and spreads), plastic carrier bags, crisp packets and cling film cannot be recycled via the kerbside economically. If too much contamination caused by unsuitable or unsafe items is collected via kerbside recycling, then this means that the whole vehicle load of recycling will be sent directly to landfill or incineration at a high cost.

Contamination is normally a problem if recyclables are collected in wheelie bins, as the worker can only look at the top; there may be contamination 'hidden' at the bottom. Councils that use many bags and boxes (Edinburgh) suffer from less contamination but are complicated and the loose paper and cardboard, and plastic recycling bags are blown around, and paper can become wet or contaminated with food residue, dirt, oil or grease.

Spain

Basque Country

In the province of Gipuzkoa, this system is implanted in many towns as Usurbil, Hernani, Oiartzun, Antzuola, Legorreta, Itsasondo, Zaldibia, Anoeta, Alegia, Irura, Zizurkil, Astigarraga, Ordizia, Oñati and Lezo, where the common used name in Basque is "atez-atekoa", which means door-by-door. Due to the big success in these towns, with more than 80% of the waste recycled, 34 towns in Gipuzkoa are considering setting this system up in 2013, like Arrasate, Bergara, Aretxabaleta, Eskoriatza, Legazpi, Tolosa or Pasaia.

The "atez-ate" system consists in hanging each kind of rubbish in a hanger outside the house a certain day or days in a week. For example, in Hernani, they have three days to hang their organic rubbish, two days for plastics and metallics, one for paper and one for rejects residuals.

This system started in the town of Usurbil in the year 2009, due to the incinerator of the region of Gipuzkoa which was going to be built in this town, exactly in the neighborhood of Zubieta. Three years after, the construction of the incinerator was stopped by the government of the region, suggesting that the incinerator was a source of contamination and the high cost of the building.

Criticism

This type of collection service is subject to criticism:

- The large (wheelie bin) container encourages the "out of sight" rubbish mentality and invites more rubbish to be disposed of.
- The bins and collection trucks are not suited to narrow roads or houses with steep driveways or steps.
- They lock local authorities into capital intensive equipment programmes and multi-national providers.
- Co-mingled recyclables are sometimes not being successfully managed by automated sorting stations and the rates of diversion are low. In some cases, this results in mountains of unsorted recyclables.
- In the UK especially, some councils are sending out at least 4 large bins - residents of smaller houses with no gardens have little space to put them; not everybody lives in a house, some live in blocks of apartments.
- Many use small plastic boxes, bags and lockable outdoor food waste 'caddies' which get blown around and lost, bad for recycling participation.

WASTE CONTAINER

A waste container is a container for temporarily storing waste, and is usually made out of metal or plastic. Some common terms are dustbin, garbage can, and trash can. The words "rubbish", "basket" and "bin" are more common in British English usage; "trash" and "can" are more common in American English usage. "Garbage" may refer to food waste specifically (when distinguished from "trash") or to municipal solid waste in general. In 1875, the first household rubbish bins were introduced in Britain to create a regulated system of collection.

Public waste container.

Public Waste Collection

Art on waste containers.

In many cities and towns, there is a public waste collection service which regularly collects household waste from outside buildings etc. This will be loaded into a garbage truck and driven to a landfill, incinerator or crush facility to be disposed of. Household waste containers are usually either:

- Trash cans, receptacles made of metal or plastic.
- Wheelie bins, light, mobile plastic bins.

In some areas, each household has multiple bins for different categories of rubbish (usually represented by colours) depending on its suitability for recycling.

Commercial roadside waste containers are often larger dumpsters or skips.

Public Litter Bins

Public areas, such as parks, often have litter bins placed to improve the social environment by encouraging people not to litter. Such bins in outdoor locations or other busy public areas are usually mounted to the ground or wall to discourage theft, and reduce vandalism, and to improve their appearance are sometimes deliberately artistic or cute. In dense urban areas, trash is stored underground below the receptacle. Some trash cans have plastic bags to help contain liquids.

Metaphors

International symbol "Tidyman" used on packaging to remind people to dispose of it in a bin instead of littering.

The term "garbage can" is also used for a model of decision making, the "Garbage Can Model" of decision making. It is concerned with cases of decision making in great aggregate uncertainty which can cause decisions to arise that from a distant point of view might seem irrational.

A "trash can" metaphor is often used in computer operating system desktop environments as a place files can be moved for deletion.

In a workplace setting, a bin may be euphemistically called "the circular file", "the round file" or "the janitor's file". Whereas useful documents are filed in a filing cabinet, which is rectangular, junk mail and other worthless items are "filed" in the bin, which is often round.

The term "waste basket" is occasionally used in taxonomy to refer to groupings that are based on some non-genetic criterion (e.g., the proposed order Insectivora is considered a "wastebasket taxon", as it groups small mammals whose chief link is that they eat insects), and the Nilo-Saharan language family is sometimes called "Greenberg's wastebasket", as it was a grouping made by him to fit the languages of Africa that did not fall into the other groups, Afroasiatic, Niger–Congo, and Khoisan.

DISPOSAL METHODS

We observe heaps of garbage lying along the roads while passing through a highway. Open dumping is the most common method of waste disposal in India. The trash heaps are usually open to the environment or seldom have sparse covering and sometimes have partial burning due to the heat generated inside.

Waste Proposal

Sometimes, wastes are illegally dumped into rivers and canals or used to fill land depressions without proper consultations. These practices cause a lot of problems in the long run, such as degrading the soil quality, causing suffocation and death of animals inhabiting that area. To prevent these, we must adopt proper waste disposal methods.

Garbage accumulation has never been much of a concern in the past, but due to globalization and industrialization, there is a need for a more efficient waste disposal method apart from the traditional methods. Following are some of the methods that are used today.

Landfill

In this process, the waste that cannot be reused or recycled separated out and spread as a thin layer in some low-lying areas across the city. A layer of soil added after each layer of garbage. Once this process is complete, this area declared unfit for building construction for the next 20 years and is only used as a playground or a park.

Incineration

Incineration is the process of controlled combustion of garbage to reduce it to incombustible matter; ash, waste gas, and heat. The waste gasses thus generated are then treated and released into the environment. This process reduced the volume of waste by 90 percent and considered as one of the most hygienic methods of waste disposal. In some cases, the heat generated used to produce electric power.

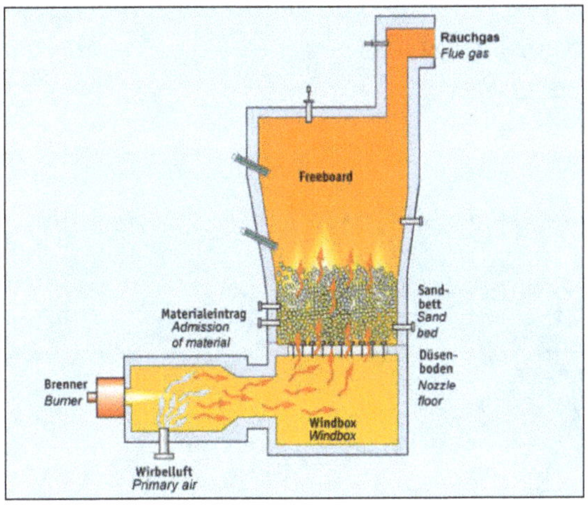

Waste Compaction

The waste materials such as cans and plastic bottles compact into blocks and send for recycling. This process prevents the oxidation of metals and reduces air space need, thus making transportation and positioning easy.

Biogas Generation

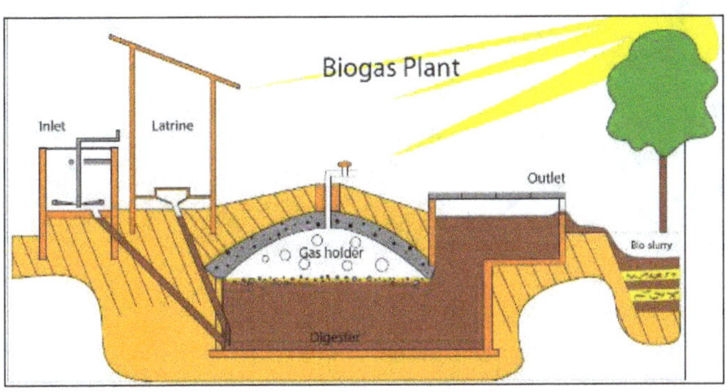

Biodegradable waste, such as food items, animal waste or industrial waste from food packaging industries send to bio-degradation plants. In bio-degradation plants, they are converted to biogas by degradation with the help of bacteria, fungi, and other biological means. Here, the organic matter serves as food for the micro-organisms. The degradation can happen aerobically (with oxygen) or anaerobically (without oxygen). Biogas is generated as a result of this process which is used as fuel, and the residue is used as manure.

Composting

All organic materials decompose with time. Food scraps, yard waste, etc. make up one of the major wastes we throw every day. Sometimes these are buried under beds of soil and left to decay under the action of microorganisms such as bacteria and fungi. This results in the formation of nutrient-rich manure. This process is termed as composting. By composting, the nutrients are replenished back to the soil. Composting also enriches the soil and increases the water retention capacity. In agriculture, it is the best alternative to chemical fertilizers.

Vermicomposting

Vermicomposting is the process of using worms for the degradation of organic matter into nutrient-rich manure. Worms take the organic matter as feed and digest them. The excretory materials or the by-products of digestion are given out by the worms makes the soil nutrient rich, thus enhancing the growth of bacteria and fungi.

References

- Waste-handling-and-transport: wastemanagement.in, Retrieved 14 February, 2019
- "Appendix 1: Kerbside Kitchen Waste Collections in New Zealand - Ministry for the Environment". Mfe.govt.nz. Retrieved 8 November 2017
- How-sort-waste-properly, en: zalajosta.lv, Retrieved 15 March, 2019
- Ministry for the Environment (December 2007). Environment New Zealand 2007. Ministry for the Environment (New Zealand). ISBN 978-0-478-30192-2. Retrieved 2008-03-27
- Tips-on-sorting-waste, en: believe.earth, Retrieved 16 April, 2019
- Shendruk, Amanada (5 August 2018). "Could NYC solve its trash problem with underground trash cans?". Quartz (publication). Retrieved 5 September 2018
- Waste-disposal, biology: byjus.com, Retrieved 17 May, 2019

5

Recycling and Re-use of Waste

The process of converting waste material into new objects and materials is known as recycling. It helps in the reduction of greenhouse gas emissions. Reusing waste materials involves the use of waste product without further transformation and without changing its shape or original nature. This chapter discusses in detail the concepts related to waste recycling and reuse such as waste exchange and remanufacturing.

WASTE RECYCLING

The definition of recycling is to pass a substance through a system that enables that substance to be reused. Waste recycling involves the collection of waste materials and the separation and clean-up of those materials. Recycling waste means that fewer new products and consumables need to be produced, saving raw materials and reducing energy consumption.

In the UK, the household and commercial sectors have relatively low recycling rates. This is in comparison to some other wastes, such as construction and demolition waste and sewage sludge. The Government is hoping to increase the amount of household waste that we recycle to 33% by 2015. Some of the materials that we can recycle include paper, plastics, metals (such as aluminium cans) and tyres.

The paper industry generates vast quantities of waste in the form of paper off-cuttings and damaged paper rolls. This paper can be put back into the pulping process and recycled. Paper recycling in the UK became popular during the 1990s. Nearly a million tonnes of paper from household waste is now recycled each year. Although paper makes up over one third of all household waste recycled, this is still no more than about 10% of the total paper consumed. In contrast, over 50% of paper waste paper produced by the newspaper industry is currently being recycled. To encourage the public to recycle waste paper, many council have arranged house to house collection schemes. Separate bins and containers are provided specifically for paper. They are collected at regular intervals and taken to be recycled. Other recycling depots for paper can be found at municipal centres and supermarkets.

Approximately 6 to 8% of UK household waste comprises of glass jars and bottles. However, the largest producers of waste glass bottles are hotels and pubs, as the vast majority of drinks are bottled. A large proportion of glass is collected in bottle banks and taken to be recycled. There are over 20,000 bottle banks in the UK, and they are mainly found in car parks and at supermarkets. There are usually three bottle banks, one for each colour of glass: clear, green and brown. The UK currently recycles about one third of its glass. This is far behind glass recycling rates in other European countries. Switzerland and the Netherlands for example have recycling rates as high as 80%.

Plastics make up a large amount of waste, since they are available in numerous forms. There are two main types of plastic: thermoplastics, which are the most common; and thermosetts. Thermoplastics melt when heated and can therefore be remoulded. This enables thermoplastics to be recycled relatively easily. In Western Europe the largest amounts of plastic occur in the form of packaging. Plastic waste tends to be sorted by hand, either at a materials recycling facility or the householder can separate it. This may then be taken to a plastic recycling point or collected by the council. The UK produces approximately about 4.5 million tonnes of plastic waste each year. Most of this waste arises from packaging. The UK has a plastics recycling rate of only 3%. In Germany the recycling rate for plastic is 70%.

The UK has a recycling rate of approximately 60% for iron and steel. Most of this waste comes from scrap vehicles, cooker, fridges and other kitchen appliances. It is estimated that the metal content of household waste is between 5 and 10%. It is mainly made up of aluminium drinks cans and tin-plated steel food cans. Aluminium recycling is widely established in the UK. It is an expensive metal and can therefore produce high incomes for recycling schemes. Copper, zinc and lead are also recycled in the UK. At present, over a third of aluminium drinks cans are recycled. Some other countries have very high recycling figures for aluminium drinks cans. The USA and Australia for example, recycle nearly two thirds.

Every year in the UK between 25 and 30 million scrap tyres are generated. Approximately 21% of these tyres are retreaded and reused. The old tread is ground off the tyre and replaced with a new tread. However, about half of all used tyres are dumped in landfill sites throughout the country. Other tyres may be incinerated.

Legislation

Supply

For a recycling program to work, having a large, stable supply of recyclable material is crucial. Three legislative options have been used to create such a supply: mandatory recycling collection, container deposit legislation, and refuse bans. Mandatory collection laws set recycling targets for cities to aim for, usually in the form that a certain percentage of a material must be diverted from the city's waste stream by a target date. The city is then responsible for working to meet this target.

Container deposit legislation involves offering a refund for the return of certain containers, typically glass, plastic, and metal. When a product in such a container is purchased, a small surcharge is added to the price. This surcharge can be reclaimed by the consumer if the container is returned to a collection point. These programs have been very successful, often resulting in an 80 percent recycling rate. Despite such good results, the shift in collection costs from local government to industry and consumers has created strong opposition to the creation of such programs in some areas. A variation on this is where the manufacturer bears responsibility for the recycling of their goods. In the European Union, the WEEE Directive requires producers of consumer electronics to reimburse the recyclers' costs.

An alternative way to increas the supply of recyclates is to ban the disposal of certain materials as waste, often including used oil, old batteries, tires, and garden waste. One aim of this method is to create a viable economy for proper disposal of banned products. Care must be taken that enough of these recycling services exist, or such bans simply lead to increased illegal dumping.

Government-mandated Demand

Legislation has also been used to increase and maintain a demand for recycled materials. Four methods of such legislation exist: minimum recycled content mandates, utilization rates, procurement policies, and recycled product labeling.

Both minimum recycled content mandates and utilization rates increase demand directly by forcing manufacturers to include recycling in their operations. Content mandates specify that a certain percentage of a new product must consist of recycled material. Utilization rates are a more flexible option: industries are permitted to meet the recycling targets at any point of their operation or even contract recycling out in exchange for tradeable credits. Opponents to both of these methods point to the large increase in reporting requirements they impose, and claim that they rob the industry of necessary flexibility.

Governments have used their own purchasing power to increase recycling demand through what are called "procurement policies". These policies are either "set-asides", which reserve a certain amount of spending solely towards recycled products, or "price preference" programs which provide a larger budget when recycled items are purchased. Additional regulations can target specific cases: in the United States, for example, the Environmental Protection Agency mandates the purchase of oil, paper, tires and building insulation from recycled or re-refined sources whenever possible.

The final government regulation towards increased demand is recycled product labeling. When producers are required to label their packaging with amount of recycled material in the product (including the packaging), consumers are better able to make educated choices. Consumers with sufficient buying power can then choose more environmentally conscious options, prompt producers to increase the amount of recycled material in their products, and indirectly increase demand. Standardized recycling labeling can also have a positive effect on supply of recyclates if the labeling includes information on how and where the product can be recycled.

Quality of Recyclate

The quality of recyclates is recognized as one of the principal challenges that needs to be addressed for the success of a long-term vision of a green economy and achieving zero waste. Recyclate quality is generally referring to how much of the raw material is made up of target material compared to the amount of non-target material and other non-recyclable material. Only target material is likely to be recycled, so a higher amount of non-target and non-recyclable material will reduce the quantity of recycling product. A high proportion of non-target and non-recyclable material can make it more difficult for re-processors to achieve "high-quality" recycling. If the recyclate is of poor quality, it is more likely to end up being down-cycled or, in more extreme cases, sent to other recovery options or landfilled. For example, to facilitate the re-manufacturing of clear glass products there are tight restrictions for colored glass going into the re-melt process.

The quality of recyclate not only supports high-quality recycling, but it can also deliver significant environmental benefits by reducing, reusing and keeping products out of landfills. High-quality recycling can help support growth in the economy by maximizing the economic value of the waste material collected. Higher income levels from the sale of quality recyclates can return value which can be significant to local governments, households, and businesses. Pursuing high-quality recycling can also provide consumer and business confidence in the waste and resource management sector and may encourage investment in that sector.

There are many actions along the recycling supply chain that can influence and affect the material quality of recyclate. It begins with the waste producers who place non-target and non-recyclable wastes in recycling collection. This can affect the quality of final recyclate streams or require further efforts to discard those materials at later stages in the recycling process. The different collection systems can result in different levels of contamination. Depending on which materials are collected together, extra effort is required to sort this material back into separate streams and can significantly reduce the quality of the final product. Transportation and the compaction of materials can make it more difficult to separate material back into separate waste streams. Sorting facilities are not one hundred per cent effective in separating materials, despite improvements in technology and quality recyclate which can see a loss in recyclate quality. The storage of materials outside, where the product can become wet, can cause problems for re-processors. Reprocessing facilities may require further sorting steps to further reduce the amount of non-target and non-recyclable material. Each action along the recycling path plays a part in the quality of recyclate.

Quality Recyclate Action Plan

The Recyclate Quality Action Plan of Scotland sets out a number of proposed actions that the Scottish Government would like to take forward in order to drive up the quality of the materials being collected for recycling and sorted at materials recovery facilities before being exported or sold on to the reprocessing market.

The plan's objectives are to:

- Drive up the quality of recyclate.
- Deliver greater transparency about the quality of recyclate.
- Provide help to those contracting with materials recycling facilities to identify what is required of them.
- Ensure compliance with the Waste (Scotland) regulations 2012.
- Stimulate a household market for quality recyclate.
- Address and reduce issues surrounding the Waste Shipment Regulations.

The plan focuses on three key areas, with fourteen actions which were identified to increase the quality of materials collected, sorted and presented to the processing market in Scotland.

The three areas of focus are:

- Collection systems and input contamination.

- Sorting facilities – material sampling and transparency.
- Material quality benchmarking and standards.

Recycling Consumer Waste

Collection

A number of different systems have been implemented to collect recyclates from the general waste stream. These systems lie along the spectrum of trade-off between public convenience and government ease and expense. The three main categories of collection are "drop-off centers", "buy-back centers", and "curbside collection".

Curbside Collection

A recycling truck collecting the contents of a recycling bin
Emptying of segregated rubbish containers.

Curbside collection encompasses many subtly different systems, which differ mostly on where in the process the recyclates are sorted and cleaned. The main categories are mixed waste collection, commingled recyclables, and source separation. A waste collection vehicle generally picks up the waste.

At one end of the spectrum is mixed waste collection, in which all recyclates are collected mixed in with the rest of the waste, and the desired material is then sorted out and cleaned at a central sorting facility. This results in a large amount of recyclable waste, paper especially, being too soiled to reprocess, but has advantages as well: the city need not pay for a separate collection of recyclates and no public education is needed. Any changes to which materials are recyclable is easy to accommodate as all sorting happens in a central location.

In a commingled or single-stream system, all recyclables for collection are mixed but kept separate from other waste. This greatly reduces the need for post-collection cleaning but does require public education on what materials are recyclable.

Source Separation

Source separation is the other extreme, where each material is cleaned and sorted prior to collection. This method requires the least post-collection sorting and produces the purest recyclates, but incurs additional operating costs for collection of each separate material. An extensive public education program is also required, which must be successful if recyclate contamination is to be avoided. In Oregon, USA, its environmental authority Oregon DEQ surveyed multi-family property managers and about half of them reported problems including contamination of recyclables due to trespassers such as transients gaining access to the collection areas.

Source separation used to be the preferred method due to the high sorting costs incurred by commingled (mixed waste) collection. However, advances in sorting technology have lowered this overhead substantially. Many areas which had developed source separation programs have since switched to what is called co-mingled collection.

Buy-back Centers

Reverse vending machine.

Buy-back centers differ in that the cleaned recyclates are purchased, thus providing a clear incentive for use and creating a stable supply. The post-processed material can then be sold. If this is profitable, this conserves the emission of greenhouse gases; if unprofitable, it increases the emission of greenhouse gasses. Government subsidies are necessary to make buy-back centres a viable enterprise. In 1993, according to the U.S. National Waste & Recycling Association, it costs on average \$50 to process a ton of material, which can be resold for \$30.

In the US, the value per ton of mixed recyclables was $180 in 2011, $80 in 2015, and $100 in 2017. In 2017, glass is essentially valueless, because of the low cost of sand, its major component; low oil costs thwarts plastic recycling. In 2017, Napa, California was reimbursed about 20% of its costs in recycling.

Drop-off Centers

Drop-off centers require the waste producer to carry the recyclates to a central location, either an installed or mobile collection station or the reprocessing plant itself. They are the easiest type of collection to establish but suffer from low and unpredictable throughput.

Distributed Recycling

For some waste materials such as plastic, recent technical devices called recyclebots enable a form of distributed recycling. Preliminary life-cycle analysis (LCA) indicates that such distributed recycling of HDPE to make filament of 3D printers in rural regions is energetically favorable to either using virgin resin or conventional recycling processes because of reductions in transportation energy.

Sorting

Once commingled recyclates are collected and delivered to a central collection facility, the different types of materials must be sorted. This is done in a series of stages, many of which involve automated processes such that a truckload of material can be fully sorted in less than an hour. Some plants can now sort the materials automatically, known as single-stream recycling. In plants, a variety of materials is sorted such as paper, different types of plastics, glass, metals, food scraps, and most types of batteries. A 30 percent increase in recycling rates has been seen in the areas where these plants exist.

Initially, the commingled recyclates are removed from the collection vehicle and placed on a conveyor belt spread out in a single layer. Large pieces of corrugated fiberboard and plastic bags are removed by hand at this stage, as they can cause later machinery to jam.

Early sorting of recyclable materials: glass and plastic bottles.

Next, automated machinery such as disk screens and air classifiers separate the recyclates by weight, splitting lighter paper and plastic from heavier glass and metal. Cardboard is removed from the mixed paper and the most common types of plastic, PET (number 1) and HDPE (number

2), are collected. This separation is usually done by hand but has become automated in some sorting centers: a spectroscopic scanner is used to differentiate between different types of paper and plastic based on the absorbed wavelengths, and subsequently divert each material into the proper collection channel.

Strong magnets are used to separate out ferrous metals, such as iron, steel, and tin cans. Non-ferrous metals are ejected by magnetic eddy currents in which a rotating magnetic field induces an electric current around the aluminum cans, which in turn creates a magnetic eddy current inside the cans. This magnetic eddy current is repulsed by a large magnetic field, and the cans are ejected from the rest of the recyclate stream.

A recycling point with separate containers for paper, plastics, and differently colored glass.

Finally, glass is sorted according to its color: brown, amber, green, or clear. It may either be sorted by hand, or via an automated machine that uses colored filters to detect different colors. Glass fragments smaller than 10 millimetres (0.39 in) across cannot be sorted automatically, and are mixed together as "glass fines".

This process of recycling as well as reusing the recycled material has proven advantageous because it reduces amount of waste sent to landfills, conserves natural resources, saves energy, reduces greenhouse gas emissions, and helps create new jobs. Recycled materials can also be converted into new products that can be consumed again, such as paper, plastic, and glass.

The City and County of San Francisco's Department of the Environment is attempting to achieve a citywide goal of generating zero waste by 2020. San Francisco's refuse hauler, Recology, operates an effective recyclables sorting facility which helped the city reach a record-breaking diversion rate of 80%.

Recycling Loops

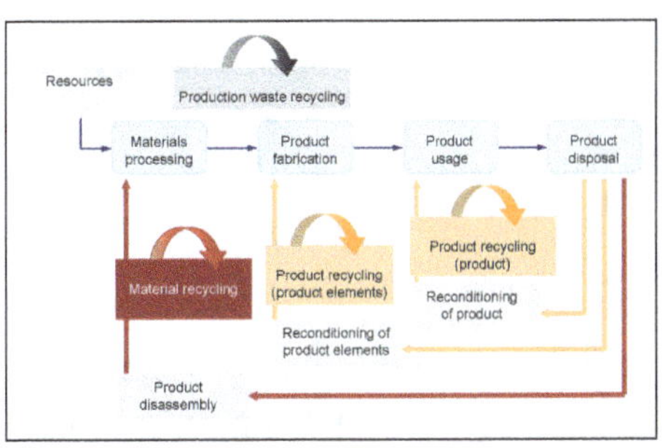

Loops for production-waste, product and material recycling.

The (ideal) recycling process can be differentiated into three loops, one for manufacture (production-waste recycling) and two for disposal of the product (product and material recycling).

The product's manufacturing phase, which consists of material processing and fabrication, forms the production-waste recycling loop. Industrial waste materials are fed back into, and reused in, the same production process.

The product's disposal process requires two recycling loops: product recycling and material recycling. The product or product parts are reused in the product recycling phase. This happens in one of two ways: the product is used retaining the product functionality ("reuse") or the product continues to be used but with altered functionality ("further use"). The product design is unmodified, or only slightly modified, in both scenarios.

Product disassembly requires material recycling where product materials are recovered and recycled. Ideally, the materials are processed so they can flow back into the production process.

Cost–benefit Analysis

Environmental effects of recycling		
Material	Energy savings	Air pollution savings
Aluminium	95%	95%
Cardboard	24%	—
Glass	5–30%	20%
Paper	40%	73%
Plastics	70%	—
Steel	60%	—

There is some debate over whether recycling is economically efficient. According to a Natural Resources Defense Council study, waste collection and landfill disposal creates less than one job per 1,000 tons of waste material managed; in contrast, the collection, processing, and manufacturing of recycled materials creates 6–13 or more jobs per 1,000 tons. However, the cost effectiveness of creating the additional jobs remains unproven. According to the U.S. Recycling Economic Informational Study, there are over 50,000 recycling establishments that have created over a million jobs in the US.

Two years after New York City declared that implementing recycling programs would be "a drain on the city", New York City leaders realized that an efficient recycling system could save the city over $20 million. Municipalities often see fiscal benefits from implementing recycling programs, largely due to the reduced landfill costs. A study conducted by the Technical University of Denmark according to the Economist found that in 83 percent of cases, recycling is the most efficient method to dispose of household waste. However, a 2004 assessment by the Danish Environmental Assessment Institute concluded that incineration was the most effective method for disposing of drink containers, even aluminium ones.

Fiscal efficiency is separate from economic efficiency. Economic analysis of recycling does not include what economists call externalities, which are unpriced costs and benefits that accrue to individuals outside of private transactions. Examples include: decreased air pollution and greenhouse gases from incineration, reduced hazardous waste leaching from landfills, reduced energy consumption, and reduced waste and resource consumption, which leads to a reduction in environmentally damaging mining and timber activity. About 4,000 minerals are known, of which only a few hundred are relatively common. Known reserves of phosphorus will be exhausted within the next 100 years at current rates of usage. Without mechanisms such as taxes or subsidies to internalize externalities, businesses may ignore them despite the costs imposed on society. To make such nonfiscal benefits economically relevant, advocates have pushed for legislative action to increase the demand for recycled materials. The United States Environmental Protection Agency (EPA) has concluded in favor of recycling, saying that recycling efforts reduced the country's carbon emissions by a net 49 million metric tonnes in 2005. In the United Kingdom, the Waste and Resources Action Programme stated that Great Britain's recycling efforts reduce CO_2 emissions by 10–15 million tonnes a year. Recycling is more efficient in densely populated areas, as there are economies of scale involved.

Wrecked automobiles gathered for smelting.

Certain requirements must be met for recycling to be economically feasible and environmentally effective. These include an adequate source of recyclates, a system to extract those recyclates from the waste stream, a nearby factory capable of reprocessing the recyclates, and a potential demand for the recycled products. These last two requirements are often overlooked—without both an industrial market for production using the collected materials and a consumer market for the manufactured goods, recycling is incomplete and in fact only "collection".

Free-market economist Julian Simon remarked "There are three ways society can organize waste disposal: (a) commanding, (b) guiding by tax and subsidy, and (c) leaving it to the individual and the market". These principles appear to divide economic thinkers today.

Frank Ackerman favours a high level of government intervention to provide recycling services. He believes that recycling's benefit cannot be effectively quantified by traditional laissez-faire economics. Allen Hershkowitz supports intervention, saying that it is a public service equal to education and policing. He argues that manufacturers should shoulder more of the burden of waste disposal.

Paul Calcott and Margaret Walls advocate the second option. A deposit refund scheme and a small refuse charge would encourage recycling but not at the expense of fly-tipping. Thomas C. Kinnaman concludes that a landfill tax would force consumers, companies and councils to recycle more.

Most free-market thinkers detest subsidy and intervention because they waste resources. Terry Anderson and Donald Leal think that all recycling programmes should be privately operated, and therefore would only operate if the money saved by recycling exceeds its costs. Daniel K. Benjamin argues that it wastes people's resources and lowers the wealth of a population.

The National Waste & Recycling Association (NWRA) then again reported in May 2015, that recycling and waste made a $6.7 billion economic impact in Ohio, U.S., and employed 14,000 people.

Trade in Recyclates

Certain countries trade in unprocessed recyclates. Some have complained that the ultimate fate of recyclates sold to another country is unknown and they may end up in landfills instead of being reprocessed. According to one report, in America, 50–80 percent of computers destined for recycling are actually not recycled. There are reports of illegal-waste imports to China being dismantled and recycled solely for monetary gain, without consideration for workers' health or environmental damage. Although the Chinese government has banned these practices, it has not been able to eradicate them. In 2008, the prices of recyclable waste plummeted before rebounding in 2009. Cardboard averaged about £53/tonne from 2004 to 2008, dropped to £19/tonne, and then went up to £59/tonne in May 2009. PET plastic averaged about £156/tonne, dropped to £75/tonne and then moved up to £195/tonne in May 2009.

Certain regions have difficulty using or exporting as much of a material as they recycle. This problem is most prevalent with glass: both Britain and the U.S. import large quantities of wine bottled in green glass. Though much of this glass is sent to be recycled, outside the American Midwest there is not enough wine production to use all of the reprocessed material. The extra must be downcycled into building materials or re-inserted into the regular waste stream.

Similarly, the northwestern United States has difficulty finding markets for recycled newspaper, given the large number of pulp mills in the region as well as the proximity to Asian markets. In other areas of the U.S., however, demand for used newsprint has seen wide fluctuation.

In some U.S. states, a program called RecycleBank pays people to recycle, receiving money from local municipalities for the reduction in landfill space which must be purchased. It uses a single stream process in which all material is automatically sorted.

Criticisms and Responses

Critics dispute the net economic and environmental benefits of recycling over its costs, and suggest that proponents of recycling often make matters worse and suffer from confirmation bias. Specifically, critics argue that the costs and energy used in collection and transportation detract from (and outweigh) the costs and energy saved in the production process; also that the jobs produced by the recycling industry can be a poor trade for the jobs lost in logging, mining, and other industries associated with production; and that materials such as paper pulp can only be recycled a few times before material degradation prevents further recycling.

Much of the difficulty inherent in recycling comes from the fact that most products are not designed with recycling in mind. The concept of sustainable design aims to solve this problem, and was laid out in the book Cradle to Cradle: Remaking the Way We Make Things by architect William

McDonough and chemist Michael Braungart. They suggest that every product (and all packaging it requires) should have a complete "closed-loop" cycle mapped out for each component—a way in which every component will either return to the natural ecosystem through biodegradation or be recycled indefinitely.

> Complete recycling is impossible from a practical standpoint. In summary, substitution and recycling strategies only delay the depletion of non-renewable stocks and therefore may buy time in the transition to true or strong sustainability, which ultimately is only guaranteed in an economy based on renewable resources.

While recycling diverts waste from entering directly into landfill sites, current recycling misses the dissipative components. Complete recycling is impracticable as highly dispersed wastes become so diluted that the energy needed for their recovery becomes increasingly excessive. "For example, how will it ever be possible to recycle the numerous chlorinated organic hydrocarbons that have bioaccumulated in animal and human tissues across the globe, the copper dispersed in fungicides, the lead in widely applied paints, or the zinc oxides present in the finely dispersed rubber powder that is abraded from automobile tires?"

As with environmental economics, care must be taken to ensure a complete view of the costs and benefits involved. For example, paperboard packaging for food products is more easily recycled than most plastic, but is heavier to ship and may result in more waste from spoilage.

Energy and Material Flows

Bales of crushed steel ready for transport to the smelter.

The amount of energy saved through recycling depends upon the material being recycled and the type of energy accounting that is used. Correct accounting for this saved energy can be accomplished with life-cycle analysis using real energy values, and in addition, exergy, which is a measure of useful energy can be used. In general, it takes far less energy to produce a unit mass of recycled materials than it does to make the same mass of virgin materials.

Some scholars use emergy (spelled with an m) analysis, for example, budgets for the amount of energy of one kind (exergy) that is required to make or transform things into another kind of product or service. Emergy calculations take into account economics which can alter pure physics-based results. Using emergy life-cycle analysis researchers have concluded that materials with large refining costs have the greatest potential for high recycle benefits. Moreover, the highest emergy efficiency accrues from systems geared toward material recycling, where materials are engineered to recycle back into their original form and purpose, followed by adaptive reuse systems where the materials are recycled into a different kind of product, and then by-product reuse systems where parts of the products are used to make an entirely different product.

The Energy Information Administration (EIA) states on its website that "a paper mill uses 40 percent less energy to make paper from recycled paper than it does to make paper from fresh lumber." Some critics argue that it takes more energy to produce recycled products than it does to dispose of them in traditional landfill methods, since the curbside collection of recyclables often requires a second waste truck. However, recycling proponents point out that a second timber or logging truck is eliminated when paper is collected for recycling, so the net energy consumption is the same. An emergy life-cycle analysis on recycling revealed that fly ash, aluminum, recycled concrete aggregate, recycled plastic, and steel yield higher efficiency ratios, whereas the recycling of lumber generates the lowest recycle benefit ratio. Hence, the specific nature of the recycling process, the methods used to analyse the process, and the products involved affect the energy savings budgets.

It is difficult to determine the amount of energy consumed or produced in waste disposal processes in broader ecological terms, where causal relations dissipate into complex networks of material and energy flow. For example, "cities do not follow all the strategies of ecosystem development. Biogeochemical paths become fairly straight relative to wild ecosystems, with very reduced recycling, resulting in large flows of waste and low total energy efficiencies. By contrast, in wild ecosystems, one population's wastes are another population's resources, and succession results in efficient exploitation of available resources. However, even modernized cities may still be in the earliest stages of a succession that may take centuries or millennia to complete." How much energy is used in recycling also depends on the type of material being recycled and the process used to do so. Aluminium is generally agreed to use far less energy when recycled rather than being produced from scratch. The EPA states that "recycling aluminum cans, for example, saves 95 percent of the energy required to make the same amount of aluminum from its virgin source, bauxite." In 2009, more than half of all aluminium cans produced came from recycled aluminium.

> Every year, millions of tons of materials are being exploited from the earth's crust, and processed into consumer and capital goods. After decades to centuries, most of these materials are "lost". With the exception of some pieces of art or religious relics, they are no longer engaged in the consumption process. Where are they? Recycling is only an intermediate solution for such materials, although it does prolong the residence time in the anthroposphere. For thermodynamic reasons, however, recycling cannot prevent the final need for an ultimate sink.

Economist Steven Landsburg has suggested that the sole benefit of reducing landfill space is trumped by the energy needed and resulting pollution from the recycling process. Others, however, have calculated through life-cycle assessment that producing recycled paper uses less energy and water than harvesting, pulping, processing, and transporting virgin trees. When less recycled paper is used, additional energy is needed to create and maintain farmed forests until these forests are as self-sustainable as virgin forests.

Other studies have shown that recycling in itself is inefficient to perform the "decoupling" of economic development from the depletion of non-renewable raw materials that is necessary for sustainable development. The international transportation or recycle material flows through "different trade networks of the three countries result in different flows, decay rates, and potential recycling returns". As global consumption of a natural resources grows, their depletion is inevitable. The best recycling can do is to delay; complete closure of material loops to achieve 100 percent recycling of nonrenewables is impossible as micro-trace materials dissipate into the environment causing severe damage to the planet's ecosystems. Historically, this was identified as the metabolic rift by Karl Marx, who identified the unequal exchange rate between energy and nutrients flowing from rural areas to feed urban cities that create effluent wastes degrading the planet's ecological capital, such as loss in soil nutrient production. Energy conservation also leads to what is known as Jevon's paradox, where improvements in energy efficiency lowers the cost of production and leads to a rebound effect where rates of consumption and economic growth increases.

This shop in New York only sells items recycled from demolished buildings.

Costs

The amount of money actually saved through recycling depends on the efficiency of the recycling program used to do it. The Institute for Local Self-Reliance argues that the cost of recycling depends on various factors, such as landfill fees and the amount of disposal that the community recycles. It states that communities begin to save money when they treat recycling as a replacement for their traditional waste system rather than an add-on to it and by "redesigning their collection schedules and trucks".

In some cases, the cost of recyclable materials also exceeds the cost of raw materials. Virgin plastic resin costs 40 percent less than recycled resin. Additionally, a United States Environmental Protection Agency (EPA) study that tracked the price of clear glass from 15 July to 2 August 1991, found that the average cost per ton ranged from $40 to $60 while a USGS report shows that the cost per ton of raw silica sand from years 1993 to 1997 fell between $17.33 and $18.10.

Comparing the market cost of recyclable material with the cost of new raw materials ignores economic externalities–the costs that are currently not counted by the market. Creating a new piece of plastic, for instance, may cause more pollution and be less sustainable than recycling a similar piece of plastic, but these factors will not be counted in market cost. A life cycle assessment can be used to determine the levels of externalities and decide whether the recycling may be worthwhile despite unfavorable market costs. Alternatively, legal means (such as a carbon tax) can be used to bring externalities into the market, so that the market cost of the material becomes close to the true cost.

Working Conditions

Some people earn their living by collecting and sorting garbage and selling them for recycling.

The recycling of waste electrical and electronic equipment in India and China generates a significant amount of pollution. Informal recycling in an underground economy of these countries has generated an environmental and health disaster. High levels of lead (Pb), polybrominated diphenylethers (PBDEs), polychlorinated dioxins and furans, as well as polybrominated dioxins and furans (PCDD/Fs and PBDD/Fs), concentrated in the air, bottom ash, dust, soil, water, and sediments in areas surrounding recycling sites.

Environmental Impact

Economist Steven Landsburg, author of a paper entitled "Why I Am Not an Environmentalist", claimed that paper recycling actually reduces tree populations. He argues that because paper companies have incentives to replenish their forests, large demands for paper lead to large forests while reduced demand for paper leads to fewer "farmed" forests.

A metal scrap worker is pictured burning insulated copper wires for copper recovery.

When foresting companies cut down trees, more are planted in their place; however, such "farmed" forests are inferior to virgin forests in several ways. Farmed forests are not able to fix the soil as quickly as virgin forests, causing widespread soil erosion and often requiring large amounts of fertilizer to maintain while containing little tree and wild-life biodiversity compared to virgin forests.

Also, the new trees planted are not as big as the trees that were cut down, and the argument that there will be "more trees" is not compelling to forestry advocates when they are counting saplings.

In particular, wood from tropical rainforests is rarely harvested for paper because of their heterogeneity. According to the United Nations Framework Convention on Climate Change secretariat, the overwhelming direct cause of deforestation is subsistence farming (48% of deforestation) and commercial agriculture (32%), which is linked to food, not paper production.

Possible Income Loss and Social Costs

In some countries, recycling is performed by the entrepreneurial poor such as the karung guni, zabbaleen, the rag-and-bone man, waste picker, and junk man. With the creation of large recycling organizations that may be profitable, either by law or economies of scale, the poor are more likely to be driven out of the recycling and the remanufacturing market. To compensate for this loss of income, a society may need to create additional forms of societal programs to help support the poor. Like the parable of the broken window, there is a net loss to the poor and possibly the whole of a society to make recycling artificially profitable, e.g. through the law. However, in Brazil and Argentina, waste pickers/informal recyclers work alongside the authorities, in fully or semi-funded cooperatives, allowing informal recycling to be legitimized as a paid public sector job.

Because the social support of a country is likely to be less than the loss of income to the poor undertaking recycling, there is a greater chance the poor will come in conflict with the large recycling organizations. This means fewer people can decide if certain waste is more economically reusable in its current form rather than being reprocessed. Contrasted to the recycling poor, the efficiency of their recycling may actually be higher for some materials because individuals have greater control over what is considered "waste".

One labor-intensive underused waste is electronic and computer waste. Because this waste may still be functional and wanted mostly by those on lower incomes, who may sell or use it at a greater efficiency than large recyclers.

Some recycling advocates believe that laissez-faire individual-based recycling does not cover all of society's recycling needs. Thus, it does not negate the need for an organized recycling program. Local government can consider the activities of the recycling poor as contributing to property blight.

Public Participation Rates

Single-stream recycling increases public participation rates, but requires additional sorting.

Changes that have been demonstrated to increase recycling rates include:

- Single-stream recycling.
- Pay as you throw fees for trash.

"Between 1960 and 2000, the world production of plastic resins increased 25-fold, while recovery of the material remained below 5 percent." Many studies have addressed recycling behaviour and strategies to encourage community involvement in recycling programmes. It has been argued that recycling behaviour is not natural because it requires a focus and appreciation for long-term planning, whereas humans have evolved to be sensitive to short-term survival goals; and that to overcome this innate predisposition, the best solution would be to use social pressure to compel participation in recycling programmes. However, recent studies have concluded that social pressure is unviable in this context. One reason for this is that social pressure functions well in small group sizes of 50 to 150 individuals (common to nomadic hunter–gatherer peoples) but not in communities numbering in the millions, as we see today. Another reason is that individual recycling does not take place in the public view.

Recycling in Art

Uniseafish - made of recycled aluminum beer cans.

Art objects are more and more often made from recycled material.

In a study done by social psychologist Shawn Burn, it was found that personal contact with individuals within a neighborhood is the most effective way to increase recycling within a community. In his study, he had 10 block leaders talk to their neighbors and persuade them to recycle. A comparison group was sent fliers promoting recycling. It was found that the neighbors that were personally contacted by their block leaders recycled much more than the group without personal contact. As a result of this study, Shawn Burn believes that personal contact within a small group of people is an important factor in encouraging recycling. Another study done by Stuart Oskamp examines the effect of neighbors and friends on recycling. It was found in his studies that people who had friends and neighbors that recycled were much more likely to also recycle than those who didn't have friends and neighbors that recycled.

Many schools have created recycling awareness clubs in order to give young students an insight on recycling. These schools believe that the clubs actually encourage students to not only recycle at school but at home as well.

RECYCLATES

"Recyclate" is a raw material that is sent to, and processed in a waste recycling plant or materials recovery facility which will be used to form new products. The material is collected in various methods and delivered to a facility where it undergoes re-manufacturing so that it can be used in the production of new materials or products. For example, plastic bottles that are collected can be re-used and made into plastic pellets, a new product.

Quality of Recyclate

The quality of recyclates is recognized as one of the principal challenges that needs to be addressed for the success of a long-term vision of a green economy and achieving zero waste. Recyclate quality is generally referring to how much of the raw material is made up of target material compared to the amount of non-target material and other non-recyclable material. Only target material is likely to be recycled, so a higher amount of non-target and non-recyclable material will reduce the quantity of recycling product. A high proportion of non-target and non-recyclable material can make it more difficult for re-processors to achieve "high-quality" recycling. If the recyclate is of poor quality, it is more likely to end up being down-cycled or, in more extreme cases, sent to other recovery options or landfilled. For example, to facilitate the re-manufacturing of clear glass products there are tight restrictions for colored glass going into the re-melt process.

The quality of recyclate not only supports high-quality recycling, but it can also deliver significant environmental benefits by reducing, reusing and keeping products out of landfills. High-quality recycling can help support growth in the economy by maximizing the economic value of the waste material collected. Higher income levels from the sale of quality recyclates can return value which can be significant to local governments, households, and businesses. Pursuing high-quality recycling can also provide consumer and business confidence in the waste and resource management sector and may encourage investment in that sector.

There are many actions along the recycling supply chain that can influence and affect the material quality of recyclate. It begins with the waste producers who place non-target and non-recyclable wastes in recycling collection. This can affect the quality of final recyclate streams or require further efforts to discard those materials at later stages in the recycling process. The different collection systems can result in different levels of contamination. Depending on which materials are collected together, extra effort is required to sort this material back into separate streams and can significantly reduce the quality of the final product. Transportation and the compaction of materials can make it more difficult to separate material back into separate waste streams. Sorting facilities are not one hundred per cent effective in separating materials, despite improvements in technology and quality recyclate which can see a loss in recyclate quality. The storage of materials outside, where the product can become wet, can cause problems for re-processors. Reprocessing facilities may require further sorting steps to further reduce the amount of non-target and non-recyclable material. Each action along the recycling path plays a part in the quality of recyclate.

PLASTIC RECYCLING

Plastic is among the most popular and important materials used in the modern world. However, its popularity is part of the huge problem and reason why plastics should be recycled. Instead of throwing them away polluting the land and our water bodies, we can optimize the lifespan of plastics by recycling and reusing them.

Plastic recycling refers to the process of recovering waste or scrap plastic and reprocessing it into useful product. Due to the fact that plastic is non-biodegradable, it is essential that it is recycled as part of the global efforts to reducing plastic and other solid waste in the environment.

Process of Plastic Recycling

The following is a step by step process of plastic recycling:

- Collection

Plastics are available in a number of forms for example plastic containers, jars, bottles, plastic bags, packaging plastic, big industrial plastics just to mention but a few. Due to their nature and availability, there are plastic collection centres and some business people have ventured into plastic collecting business as a source of income. Tons and tons of scrap plastic are collected and sent to a collecting yard where they are then packed and transported to plastic processing plants. Unfortunately, not all countries have the capacity to recycle plastic. Very few developing countries can actually recycle plastic. This means that, plastic waste is still a major problem to some countries in the world.

- Sorting

The actual plastic recycling process starts with sorting of the different plastic items by their resin content and color. This process is also done to ensure all contaminates are eliminated. There are specially designed machines that help in sorting of the plastics according to their resin content. Then the recycling mill sorts the scrap plastic by symbols at the bottom of the plastics.

- Shredding

After sorting the plastics, the next step is to cut the plastics into tiny chunks or pieces. The plastic bottles and containers are then ground and cut into tiny pieces or flakes. The heavier and lighter plastic flakes are separated using a specially designed machine. The separation process helps in ensuring that the different plastics are not put together or mixed up in the final product. Remember that different plastics are used to make different items.

- Cleaning

After a complete separation, the flakes or chunks are then washed with detergents to remove the remaining contamination. Once the cleaning process is complete, the clean flakes are passed through specialized equipment that further separates the plastic resin types. The plastic flakes are then subjected to moderate heat to dry.

- Melting

The dry flakes are melted down. They can be melted down and molded into a new shape or they are melted down and processed into granules. The melting process is done under regulated temperatures. There is specialized equipment designed to melt down plastic without destroying them.

- Making of pellets

After the melting process, the plastic pieces are then compressed into tiny pellets known as nurdles. In this state, the plastic pellets are ready for reuse or be redesigned into new plastic products. It is important to point out that recycled plastic is hardly used to make identical plastic item or its previous form. It is in this pellet form that plastics are transported to plastic manufacturing companies to be redesigned and be used in making other useful plastic products.

Common Recycled Plastics

There are numerous and common types of recycled plastics as can be seen below:

- Polyethylene Terephthalate

This type of recycled plastic is tough, has excellent clarity, is strong and has barrier to moisture and gas. It is used in the manufacture of water, soft drinks, peanut butter and salad dressing bottles and jars.

- High Density Polyethylene

This recycled plastic is known for its excellent stiffness, resistance to moisture, strength, versatility, toughness and reduced permeability to gas. It is used in the manufacture of water, juice and milk bottles. It is also used to make retail and trash bags for households and business people.

- Polyvinyl Chloride

Abbreviated as PVC, polyvinyl chloride has a number of applications. It is versatile, can be bended easily, it is tough and strong. This recycled plastic is commonly used in the manufacture of juice bottles, PVC piping and cling films.

- Low Density Polyethylene

This is the most common type of recycled plastic. It has exceptional ease of processing; it is strong, flexible, tough, and resistant to moisture and it's easy to seal. This plastic is usually used in making frozen food bags, flexible container lids, freezable bottles.

Advantages of Recycling Plastics

Plastics should be recycled because of a number of reasons as can be seen below:

- Provision of a Sustainable Source of Raw Materials

Recycling plastics provides a sustainable source of raw materials to the manufacturing industry. Once the plastics are recycled, they are sent to manufacturing industries to be redesigned and converted into new shapes and used in different appliances.

- Reduces Environmental Problems

Since plastics are non-biodegradable, they pose a high risk to the people and the environment as a whole. They can block sewer lines, drainages and other waterways leading to blockages and unwanted pileups. When plastics are eliminated through recycling, the environment looks clean and inhabitable.

- Reduces Landfill Problems

Recycling plastics minimizes the amount of plastic being taken to the ever diminishing landfill sites. Most countries have designated areas specifically meant for burying plastics. When they are recycled, these sites will receive little plastic garbage. The remaining areas can be used for other purposes instead of dumping plastics that do not rot. These areas can be used for agriculture or for human settlement. It should be understood that human population is growing each day and land is becoming a problem. Instead of misusing the land for garbage disposal it can be used for settlement and other important economic activities.

- Consumes Less Energy

Recycling of materials including plastics requires less energy as compared to making the plastic from scratch. This saves energy and that energy can be diverted to other important things in the economy. It is therefore important to encourage plastic recycling in the manufacturing industry as it will save the economy billions of money. The process of manufacturing plastic using natural raw materials is expensive and time consuming compared to the recycling process.

- Encourages a Sustainable Lifestyle among People

Individuals who have ventured into plastic collection and recycling business will experience improved lifestyles as they will get their daily income from the business. This will in the long run improve the economy and boost the living standards of the people. So do not just sit there doing nothing, embrace plastic recycling activities and improve your economic standards.

Any sort of effort aimed at saving the environment is very important and matters a lot. Since its inception during the environmental revolution in the late 1960s, plastic recycling is one of the most encouraged solid waste management programs in the world. Prior to the push to use of plastic containers by manufacturers, products were packaged in glass, metal and paper. Therefore, in order to keep our environment clean, reduce landfills, provide a sustainable supply of plastics to manufacturers, it is important to recycle plastics.

REUSE

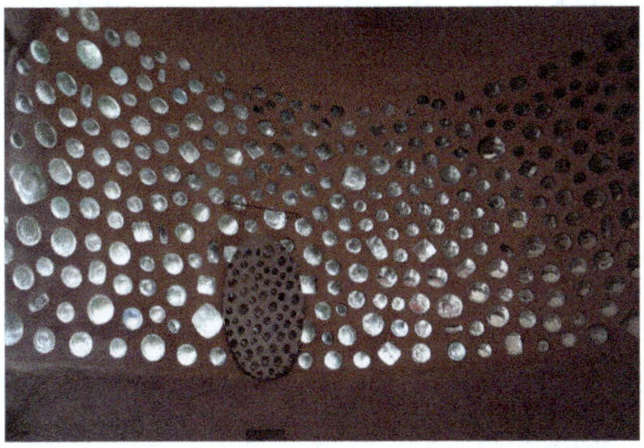

An interior bathroom wall that incorporates repurposed clear glass bottles into a bottle wall.

Reuse is the action or practice of using something again, whether for its original purpose (conventional reuse) or to fulfil a different function (creative reuse or repurposing). It should be distinguished from recycling, which is the breaking down of used items to make raw materials for the manufacture of new products. Reuse – by taking, but not reprocessing, previously used items – helps save time, money, energy and resources. In broader economic terms, it can make quality products available to people and organizations with limited means, while generating jobs and business activity that contribute to the economy.

Historically, financial motivation was one of the main drivers of reuse. In the developing world this driver can lead to very high levels of reuse, however rising wages and consequent consumer demand for the convenience of disposable products has made the reuse of low value items such as packaging uneconomic in richer countries, leading to the demise of many reuse programs. Current environmental awareness is gradually changing attitudes and regulations, such as the new packaging regulations, are gradually beginning to reverse the situation.

One example of conventional reuse is the doorstep delivery of milk in reuseable bottles; other examples include the retreading of tires and the use of returnable/reusable plastic boxes, shipping containers, instead of single-use corrugated fiberboard boxes.

Advantages and Disadvantages

Ants, being social insects, have been able to reuse rail tracks abandoned by humans for their own transportation.

Reuse has certain potential advantages:

- Energy and raw materials savings as replacing many single use products with one reusable one reduces the number that need to be manufactured.
- Reduced disposal needs and costs.
- Refurbishment can bring sophisticated, sustainable, well paid jobs to underdeveloped economies.
- Cost savings for business and consumers as a reusable product is often cheaper than the many single use products it replaces.
- Some older items were better handcrafted and appreciate in value.

Disadvantages are also apparent:

- Reuse often requires cleaning or transport, which have environmental costs.
- Some items, such as freon appliances, infant auto seats, older tube TVs and secondhand automobiles could be hazardous or less energy efficient as they continue to be used.
- Reusable products need to be more durable than single-use products, and hence require more material per item. This is particularly significant if only a small proportion of the reusable products are in fact reused.
- Sorting and preparing items for reuse takes time, which is inconvenient for consumers and costs money for businesses.
- Special skills are required to tweak the functional throughput of items when devoting them to new uses outside of their original purpose.
- Knowing the standards that legacy products conform to is required for knowing what adapters to buy for newer products to be compatible with them, even though the cost of adapters for such applications is a minor disadvantage.
- Being a rather minor disadvantage, metal that is repurposed later on can sometimes contain rust, seeing as it sometimes ages before reuse.

Addressing Issues of Repair, Reuse and Recycling

One way to address this is to increase product longevity; either by extending a product's first life or addressing issues of repair, reuse and recycling. Reusing products, and therefore extending the use of that item beyond the point where it is discarded by its first user is preferable to recycling or disposal, as this is the least energy intensive solution, although it is often overlooked.

The EU Circular Economy Package recognises the importance of extending product lifetimes and includes repair and reuse of products in its action plan to ensure products reach their optimum lifespan. If targets for reducing greenhouse gas emissions are to be reached, then reuse needs to be included as part of a whole life cycle approach.

A strong second hand market-place exists, with charity shops on most high streets, car boot(trunk) sales and online auction sites maintaining popularity and regular TV shows featuring both buying and selling at auction.

Inadequate Repair Infrastructure

However, the reuse of electrical and electronic equipment products remains low due to inadequate infrastructure, lack of repair networks and products. Local authority collection systems are better suited for handling waste than handling goods and preserving reuse potential. Retailer delivery staff are trained to handle goods carefully.

So, do we need a radical rethink on how we move unwanted, still usable items to the second-hand market place? Is there a case for changing our approach to producer responsibility and insisting that producers finance collection for reuse, and additionally, drive consumer choices for reuse, repair and remanufacture; whilst addressing the costs of recycling and disposal?

There are opportunities for producers, waste management companies and local authorities to make both repair and reuse habitual, whilst these require changes to householder behaviour change through raising awareness, they also require investment in infrastructure and logistical operations. Is it time to insist that more products are designed to have longer lifetimes?

Measuring the Impact of Reuse Metrics

Determining the balance of how the several effects of reuse interact is often best accomplished by a formal life cycle assessment. For example, research has shown that reusing a product can reduce CO_2 emissions and carbon footprint by more than 50% relative to the complete product life cycle. A relatively unknown effective way to reduce CO_2 emissions and carbon footprint is reusing products. Often the relative carbon footprint of manufacturing and the supply chain is unknown. A scientific methodology has been developed to calculate how much CO_2 emissions are reduced when buying used or second hand hardware versus new hardware, the so-called durability greener network calculator.

There are many ways of measuring the positive environmental, economic and social impact data. These include:

- Number of tons diverted from the landfill.
- Avoided disposal costs.
- Avoided purchase costs.
- Value of materials donated.
- Revenues earned.
- Number of jobs created or retained.
- Number of families/individuals/organizations assisted.

Internalized Environmental Costs

A Pigovian tax is an environmental tax: a charge on items that reflects the environmental costs of their manufacture and disposal. This makes the environmental benefit of using one reusable item instead of many disposable ones into a financial incentive. Such charges have been introduced in some countries.

Comparison to Recycling

Recycling differs from reuse in that it breaks down the item into raw materials which are then used to make new items, as opposed to reusing the intact item. As this extra processing requires energy, as a rule of thumb reuse is environmentally preferable to recycling ("reduce, reuse, recycle"), though recycling does have a significant part to play as it can often make use of items which are broken, worn out or otherwise unsuitable for reuse. However, as transport emissions are significant portion of the environmental impact of both reuse and recycling, in some cases recycling is the more prudent course as reuse can require long transport distances. A complex life cycle analysis may be required during a product's design phase to determine the efficacy of reuse, recycling, or neither, and produce accordingly.

A school being prepared for reuse as housing.

Reuse of Information

Besides physical resources, information is often reused, notably program code for the software that drives computers and the Internet, but also the documentation that explains how to use every modern device. And it is proposed as a way to improve education by assembling a great library of shareable learning objects that can be reused in learning management systems.

Software reuse grew out of the standard subroutine libraries of the 1960s. It is the main principle of today's object-oriented programming. Instead of constantly reinventing software wheels, programming languages like C++, Java, Objective-C, and others are building vast collections of reusable software objects and components.

Reuse of information has a tremendous return on investment for organizations whose documentation is translated into many languages. Translation memory systems can store text that has already been translated into dozens of languages for retrieval and reuse.

WASTE EXCHANGE

A Waste Exchange is an operation that enables materials discarded by one company to be re-used by another company. These can be scrap, production by-products, obsolete or unused raw materials, hazardous waste and recyclable products.

The potential cost savings that can be made are immense as on average businesses spend 4.5 percent of their total turnover on waste disposal and a significant percentage on raw materials. Many of these costs can be removed through exchanging materials, turning what was once a significant drain on financial resources into profit.

In general, a waste exchange system (also called a 'materials exchange' system) is an alternative to disposal of waste to landfill. Such systems enable the efficient reuse and recycle of materials, reduction in waste that has to be disposed, cost savings in raw materials and waste disposal, and enables compliance with waste laws and regulations where relevant.

Through a waste exchange system, companies can find alternatives to landfilling of unwanted materials, save on disposal costs and support community groups and schools by registering re-usable materials. Through waste exchange, users can access re-usable materials, access information about local recyclers and regional recycling services or reduce rubbish going to landfill.

A quick survey of waste exchange systems established worldwide reveals four key functions:

- Database Function

This is the most important and core function of a waste exchange system. Usually made available online as a web-based database, offers and requests for wastes are included in the database and are searchable. In some cases, advisory services and 'waste matching' services may also be included.

- Information Function

Most waste exchange systems have a vast array of general waste management information, functioning as a repository both online and offline.

- Interaction Function

Typically, this function includes the organization of seminars, workshops, training and other activities for members or users of the waste exchange system.

- Communication Function

Registered members of the waste exchange as well as the public at large, are provided services such as newsletters, email reminder services, information sharing etc.

The main function of a waste exchange system, of course is its Database function. In populating a database, the following points are usually collected:

- Material offer or request.
- Material classifications: acids, alkalis, other inorganic chemicals, solvents, other organic chemicals, oils and waxes, plastics and rubber, textiles and leather, wood and paper, metals and metal sludges, etc.
- Generic name of material and description.
- Potential or intended use.
- Packaging: bulk, drums, pallets, bales, etc.

- Present amount: in lbs., kgs, tons, gal., etc.
- Frequency: continuous supplies, variable supplies, one time supply, etc.
- Restrictions, if any (including hazardous characteristics).
- Available to interested parties: sample lab analysis, independent analysis, consultation, etc.

Some features of a typical waste exchange system:

- Waste exchanges behave as a B2B stock exchange for wastes.
- Waste exchange systems common goals are to achieve a recourse efficient and sustainable society. They achieve these aims through a number of ways - promotion and facilitation sector collaboration and partnerships across industry, government and the community, implement innovative actions and projects, actively contribute to and support relevant government policy, strategies and actions, and foster knowledge and skills through communications, and information and educational resources.
- Some of the most common reasons business cite to use a waste exchange system include (a) receiving low or no-cost materials, (b) reduction of disposal and purchase costs, (c) free-up storage space, (d) find markets for surplus materials/scrap.
- Materials exchanged through the system usually consist of business-related items in good, usable condition (e.g., business overruns, seconds, outdated inventory, empty containers and office furniture).
- Examples of wastes exchanged fall under a number of categories, including for example, boxes and packaging, chemicals, computer/electrical, construction materials, containers and pallets, metals, organic waste, paints and stains, paper/cardboard, plastic materials, textiles, wood waste etc.
- Besides their characteristics, materials are also listed as available or required along a timeline - continuously, ongoing, one-off, daily, weekly, or monthly.
- Costs of the exchange service itself is in many cases free (run by a non-profit or subsidized by local governments and business associations), though a fee may be incurred in transportation and payment for the wastes purchased.
- Waste exchange services mostly cater only to companies and other organizations, and not residential or household entities. However, many non-profit organizations sometimes take advantage in using waste exchanges.
- User-members of waste management systems include elected officials, managers, engineers, planners, policy makers, operators, and the organizations they represent - including central, regional and local government agencies, private waste management companies, equipment suppliers and service providers, engineering and consulting companies, non-profit organizations etc.
- Waste exchange systems do not deal only with wastes. Many organizations also host conferences, conduct workshops and seminars, publish newsletters, share information with its members, or provide networking opportunities for its members.

- Due to copyright, patents and IPR issues, many waste management systems have a robust confidentiality policy in place, enabling companies to exchange wastes without infringing on these rights.
- Some waste exchange systems also deal with waste management resources and waste trading tools to improve recycling and buy and sell scraps, recycled materials and products made from recycled materials.
- Some combine waste information exchange services (as a clearinghouse and repository for industry) providing information on, for example:
    - Market development.
    - Recycled products.
    - Solid and hazardous waste management.
    - Current regulations/legislation.
    - Alternative and emerging waste management technologies.
    - Trade journals and associations.
    - Technical reports.
    - Waste management services and products.

REUSABLE PACKAGING

Reusable packaging is manufactured of durable materials and is specifically designed for multiple trips and extended life. A reusable package or container is "designed for reuse without impairment of its protective function." The term returnable is sometimes used interchangeably but it can also include returning packages or components for other than reuse: recycling, disposal, incineration, etc. Typically, the materials used to make returnable packaging include steel, wood, polypropylene sheets or other plastic materials.

Reusability of packaging is an important consideration of the environmental credo of "reduce, reuse, and recycle". It is also important to the movement toward more sustainable packaging. Returnable packaging is encouraged by regulators.

Shipping Containers

For many years, several types of shipping containers have been returnable and reusable. These have made most sense when a reverse logistics system is available or can be readily developed. A return, reconditioning, and reuse system can save money on the cost per shipment and can reduce the environmental footprint of the packaging.

Manufacturing, particularly the automotive industry, has used heavy-duty returnable racks for shipping hoods, fenders, engines, dashboards, etc. from suppliers to final assembly plants. The racks are then returned for the next shipment cycle.

Bulk foods, chemicals, and pharmaceuticals are often shipped in reusable and returnable containers. These need to be carefully inspected, cleaned and sanitized as part of the reuse cycle. An effective Quality Management System is necessary.

Wooden pallets are often made to be expendable, for a single shipment. Others are heavy duty and intended for multiple shipments. Some are in "pallet pools" which are used, inspected, and refurbished for extended usage. Often reusable industrial shipping containers have bar code labels or RFID chips to help identify and route the containers.

Use in the Automotive Industry

Automotive OEM manufacturers use and encourage the use of returnable packaging to move components from their vendors to their factories. The components are placed in returnable packaging and are at times and arranged in a way that facilitates movement straight to assemble lines. Such packaging replaces traditional corrugated cartons, thereby helping companies cut costs by avoiding wastage and effort required in disposing the cartons. It also helps in reducing the environmental footprint of the automotive industry.

Other advantages of using returnable packaging include avoiding damages to parts in while in transit. Parts are at times placed in specially designed receptacles for easy picking on the assembly line contributing to fewer mistakes and simpler inventory management.

A few examples of returnable packaging in automotive industry:

A steel cage used as reusable packaging.

A wooden pallet collar box.

Consumer Packaging and Containers

Several types of consumer containers have been in reuse systems. Reusable bottles for milk, soda, and beer have been part of closed-loop use-return-clean-refill-reuse cycles. Food storage containers are typically reusable. Thick plastic water bottles are promoted as an environmental improvement over thin single-use water bottles. Some plastic cups can be re-used, though most are disposable.

Home canning often uses glass mason jars which are often reused several times. Many non-food types of containers, including reusable shopping bags, and luggage, are designed to be reused by consumers. With any food packaging, proper cleaning and disinfecting between each use is critical to health.

Reuse for Other Purposes

Used packages are often reused for purposes other than their primary use. For example, a single-use plastic shopping bag might be reused as a bin bag, a household storage bag or a dog faeces bag.

Justification

Reusable packaging often costs more initially and uses more and different materials than single-use packaging. It often requires adding complexity to the distribution system. Not all packaging justifies being returnable and reusable.

A thorough cost analysis is required. This involves all of the material, labor, transport, inspection, refurbishing, cleaning, and management costs. Often these costs may be incurred by different companies with different cost structures.

The environmental costs and benefits can also be complex. The material, energy, pollution, etc. needs to be accounted for throughout the entire system. A Life Cycle Assessment offers a good methodology for this task.

REMANUFACTURING

Remanufacturing is a process where a particular product is taken apart, cleaned, repaired, and then reassembled to be used again. Remanufacturing has long been associated with expensive technical products, but the technique is spreading. C. Franke and his coauthors, writing in Omega made the point as follows: "Today, the remanufacturing of expensive, long-living investment goods, e.g., machine tools, jet fans, military equipment or automobile engines, is extended to a large number of consumer goods with short life cycles and relatively low values. Reuse is an alternative to material recycling to comply with recovery rates and quantities as well as special treatment requirements" mandated by regulatory authorities. The list today includes mobile phones, tires, furniture, laser toner cartridges, computers, and electrical equipment. Essentially any product that can be manufactured can also be remanufactured. In order for a product to be considered remanufactured, most of its components must be used, although some of them can be new if the older parts are too defective to be salvaged.

Remanufacturing thus has two underpinnings. One is economic and the other is public or governmental regulatory pressure. From an environmental view point, remanufactured good are held out of the waste stream, conserve energy and thus reduce green-house gases, and protect ground-water from potentially toxic leachates—especially important in context of electronic goods. The economic motive is obvious in the case of very massive and expensive products such as machine tools and oceangoing vessels; they can also be quite real if public participation in the return of the products in part subsidizes the costs of their return to a remanufacturing facility.

While the basic concept of remanufacturing is simple, the activity is complex. It requires that a used product be completely disassembled in order to assess its actual condition. If it is determined that remanufacturing is worthwhile, various parts of the product are cleaned, restored, repaired, and replaced. Further refinements are then performed and the product is reassembled so that it once again operates in the way it was originally intended to function. The product is then ready to be used again. Each step in this process is essential to the entire concept of remanufacturing and careful precautions must be taken to ensure that each step is carried out correctly.

Related Processes

The reuse of an object can take place after application of different kinds of processes and in various forms. The simplest form of fundamental reuse is recycling, represented by steel or aluminum beverage cans extracted from waste or separately collected which are then reintroduced into steel or aluminum furnaces scrap and may return in some other kind of form to the market.

Similar to recycling is a process of disassembly sometimes referred to as "demanufacturing"—after which the components thus obtained may be handled by recycling processes, remanufacturing methods, direct sale to end users, or by disposal. Many automobiles delivered to junk yards are demanufactured. Engines are removed and sometimes sold to remanufacturers, components pars are sold as found to individuals or repair shows, seats are removed and sold or disposed of as waste, structural components are separated and sold as scrap steel. Ship-breaking follows a similar cycle.

Certain products seen by the consumer as single entities may have the distinct roles of "container" and "content." The classical case is a returnable bottle which ends up, without its closure, at the bottling plant again to be cleaned, refilled with soda, and closed with a new cap. Toner cartridges used in laser printers are such container-content combinations, the cartridge itself designed for reuse, the toner used in printing.

In instances where the product is remanufactured, it will once again end up performing the same function it performed before after a more or less intensive remanufacturing process. To meet the "remanufactured" definition, the product must undergo some extensive process that is significantly more than "repair." A simple example of remanufacturing is the retreaded tire in which the basic inner core of the tire is retained, the remaining tread is cut off, and new rubber is applied and bonded to the core. In essence remanufactured products undergo significant processing beyond cleaning, repair, and maintenance. They are thus restored to a much higher functionality as a "used" product. Many auto parts must be remanufactured for sustained use and represent a major element of the remanufacturing industry.

Industry Size and Benefits

The Automotive Parts Remanufacturers Association (APRA), citing research conducted by Boston University, estimated that remanufacturing had sales of $52 billion in the United States and estimated volume in excess of $100 billion worldwide. In the U.S., more than 70,000 firms are active in some kind of remanufacturing. APRA also cited data from the German Fraunhofer Institute to the effect that energy savings worldwide due to remanufacturing exceed the equivalent of 10.7 million barrels of crude oil. Substantial elimination of solid waste generation and atmospheric pollution follows.

Remanufacturing and Small Businesses

Aside from environmental benefits, there are many other reasons why remanufactured goods exist. Like many good business decisions, remanufacturing simply saves money by prolonging the economic life of a product. A small business with a tight budget can save money by using remanufactured products because they often cost less (anywhere between 40 and 60 percent less) and come with warranties and extra services that guarantee their performance.

Remanufacturing is also a business opportunity for small businesses with the appropriate technical skills and equipment deployments. For example, an auto repair business can potential branch out and begin to offer remanufactured goods as part of its services, or a small business that repairs office machines may be able to gain the necessary knowledge to remanufacture related products at the same time as it conducts its normal business activities.

If a small business decides to get into the remanufacturing industry, it must first and foremost study and understand the market. Despite the recent success of remanufacturing, there is still a negative perception among consumers regarding products that contain used parts. Many consumers feel that a remanufactured product is not durable as a brand new one and may require additional maintenance in the future. This is a serious issue that must be addressed before a small business decides whether it is worth it to pursue remanufacturing as a vocation.

Like any business venture, remanufactured products must be properly marketed in order for the company producing them to ultimately succeed. Management must target consumers who will appreciate the fact that remanufactured goods are a great financial alternative to new ones, but educate them enough so that they understand they are not sacrificing quality for price. A sound warranty plan and follow-up calls that gauge the product's performance are also suggested. Like any product or service, a remanufactured product will benefit from positive word of mouth and grow into a solid business because of it.

Inexperienced remanufacturing firms must also be careful not to compete against themselves when marketing remanufactured and new goods at the same time. In addition, management must work with their own employees so that they understand the many benefits of the remanufacturing process. Many employees may be hesitant to offer remanufactured goods to their customers for fear of a potential prejudices regarding the performance of the product.

Most importantly, a small business must have the means at its disposal to locate and recover the products and resources that will be used in the remanufacturing project and ultimately perform the task at hand. Once these products are found, they must be transported to the destination where disassembly will take place. After that, they will most likely be transported to another location that specializes in reassembly. Finally, any unusable parts and products must be collected and transported to recycling centers or other places that specialize in their disposal.

There are many legal and regulatory issues that affect the remanufacturing industry that businesses must be aware of. Intellectual property and anti-trust matters; federal, state and local recycling procedures; and government economic incentives are just a few of these issues. The Remanufacturing Institute is the watchdog organization for the entire industry and they are constantly monitoring these issues and representing the views of the businesses that are involved in remanufacturing.

In addition, the federal government requires that all remanufactured goods must be labeled as such so that they cannot be passed off as new products.

RECYCLING OF INDUSTRIAL WASTE

Waste outputs for industrial materials like solvents, paints, oils and adhesives are difficult to deal with. Either laden with or exposed to chemicals hazardous to humans and the environment, the proper disposal and containment of these products and their packaging is highly regulated by the U.S. Environmental Protection Agency (EPA) and cannot typically be captured or processed by municipal waste systems. Companies, manufacturers and small businesses are thereby responsible for the private management of this waste, which typically entails linear disposal solutions like incineration, land disposal and underground injection wells.

Industrial waste materials must be disposed of properly and safely, but like all types of waste, viable waste solutions are contingent upon economics. There is little economic incentive to employ other solutions for discarded industrial waste materials in addition to the usual linear disposal, which is often quite costly to begin with. Non-linear technologies for hazardous waste management now exist, but by and large, most businesses dealing with industrial materials do not offer regenerative waste solutions for their products and packaging.

Difficult-to-recycle waste outputs like industrial sealants can be managed effectively by bringing a perspective of value to its component parts as potential inputs to new production cycles. Another example of a company that recycles its industrial waste, and makes a profit while doing it, is General Motors. Among the materials diverted from landfills through circular systems, paint sludge has been turned into plastic containers durable enough to hold Chevrolet volt and Cruze engine components, and solvents used between paint color changes have been reformulated into paints applied to plant floors. GM is a large-scale example of a manufacturer that has created a production infrastructure that makes capturing and processing their post-industrial waste revenue-friendly.

Ways to Recycle Industrial Waste

Before one understands the fundamentals of recycling, let's see why we bother with recycling in the first place! Every day, industrial establishments generate immense amounts of toxic waste that are deadly to the environment. Recycling helps reduce the emission of greenhouse gas and landfill space by a large degree. Energy is also conserved when we use the waste materials to make new stuff. Let's see some of the methods that can be used to do so:

Breaking the Waste Down

As an environment-friendly alternate to landfills, industrial composting systems are now being used increasingly. To break down biodegradable organic waste into useful products, microorganisms are used. The resultant produce from this process is also being used as an alternative fuel in some countries. Currently, composting takes place in most Western countries and is mandatory in quite a number of them.

Burn it to Make it

Some materials can be remelted to form new objects. Take plastic bottles; they can be converted into clothing-grade polyester. The problem with this method is that every subsequent time the process takes place, levels of unwanted substances increase. Industrial transformers, oil filters, and large containers are often recycled through this method.

Hot Treatment

Most waste materials contain energy content to some degree. This energy can be used for cooking, heating and generating steam, by heating the waste at high temperatures with limited oxygen availability. Using this type of thermal treatment, the energy content can be processed into alternative fuels too.

Harnessing the Harmful Gas

Recycling promotes the avoidance of landfills altogether, but since dumping waste in an abandoned area is so ridiculously inexpensive, landfills are still quite common. Thus, as part of resource recovery, the harmful landfill gas is captured and harnessed to generate fuel or electricity.

Reclaiming the Waste

In reclamation, waste is processed and useful material is extracted from it. For example, mercury can be recovered from broken thermometers. Lead can be reclaimed from batteries and paint. Many solvents that have been used can also be distilled for reuse, like acetone.

These are only some of the ways that are being used today to recycle industrial waste. Recycling non-industrial waste is a whole other area. The reason that recycling of industrial waste is usually given much more importance is that, the amount of solid and liquid waste being dumped each day by industries is flabbergasting to say the least.

References

- Waste-Recycling, Sustainability: enviropedia.org.uk, Retrieved 18 June, 2019
- "Publications – International Resource Panel". Unep.org. Archived from the original on 11 November 2012. Retrieved 7 July 2016
- Recyclingplastic: conserve-energy-future.com, Retrieved 19 July, 2019
- Mata, T M (2001). "Life cycle assessment of different reuse percentages for glass beer bottles". International Journal of Life Cycle Assessment. 6 (5): 58–63. Doi:10.1007/BF02978793
- Waste-exchange, waste, uem: gdrc.org, Retrieved 20 August, 2019
- Environmental Protection Agency Frequently Asked Questions about Recycling and Waste managementarchived 27 September 2006 at the Wayback Machine. Retrieved 18 October 2006

Permissions

All chapters in this book are published with permission under the Creative Commons Attribution Share Alike License or equivalent. Every chapter published in this book has been scrutinized by our experts. Their significance has been extensively debated. The topics covered herein carry significant information for a comprehensive understanding. They may even be implemented as practical applications or may be referred to as a beginning point for further studies.

We would like to thank the editorial team for lending their expertise to make the book truly unique. They have played a crucial role in the development of this book. Without their invaluable contributions this book wouldn't have been possible. They have made vital efforts to compile up to date information on the varied aspects of this subject to make this book a valuable addition to the collection of many professionals and students.

This book was conceptualized with the vision of imparting up-to-date and integrated information in this field. To ensure the same, a matchless editorial board was set up. Every individual on the board went through rigorous rounds of assessment to prove their worth. After which they invested a large part of their time researching and compiling the most relevant data for our readers.

The editorial board has been involved in producing this book since its inception. They have spent rigorous hours researching and exploring the diverse topics which have resulted in the successful publishing of this book. They have passed on their knowledge of decades through this book. To expedite this challenging task, the publisher supported the team at every step. A small team of assistant editors was also appointed to further simplify the editing procedure and attain best results for the readers.

Apart from the editorial board, the designing team has also invested a significant amount of their time in understanding the subject and creating the most relevant covers. They scrutinized every image to scout for the most suitable representation of the subject and create an appropriate cover for the book.

The publishing team has been an ardent support to the editorial, designing and production team. Their endless efforts to recruit the best for this project, has resulted in the accomplishment of this book. They are a veteran in the field of academics and their pool of knowledge is as vast as their experience in printing. Their expertise and guidance has proved useful at every step. Their uncompromising quality standards have made this book an exceptional effort. Their encouragement from time to time has been an inspiration for everyone.

The publisher and the editorial board hope that this book will prove to be a valuable piece of knowledge for students, practitioners and scholars across the globe.

Index

A

Activated Sludge, 18, 20

Anaerobic Digestion, 3, 16-17, 20, 82, 85, 188, 202

B

Bedrock Disposal, 144, 146-147

Bio-medical Waste, 77-78

Biodegradable Waste, 1-3, 14, 16, 190, 192, 197, 202, 207

Biofuel, 9, 23

Biogas Production, 77, 126

Biological Oxygen Demand, 5, 14

Biological Waste, 7, 60

Borosilicate Glass, 52

Brominated Flame Retardants, 19, 23,30, 36

C

Cathode Ray Tubes, 2, 25, 29

Ceramic Waste, 149-150

Chemical Oxygen Demand, 5, 110, 145

Constructed Wetlands, 6, 85, 113-115

D

Downcycling, 31, 162, 164

Dust Collection System, 33

E

E-waste, 2, 23-34, 37-41, 65, 182-183, 199

E-waste Recycling, 27-28, 31-32, 37-40, 183

Energy Recovery, 1, 67, 143, 152, 154, 160, 162, 187

Eutrophication, 6, 91

Extended Producer Responsibility, 152, 165, 174, 179-180, 182-183, 185

F

Feedlot Runoff, 102, 114

Fly Ash, 8, 42, 48, 53, 67, 222

G

Garbage Grinders, 64, 66

Geologic Disposal, 53, 149

Green Waste, 1, 3, 7-9, 59, 197-199

Greenhouse Gases, 9, 14-15, 79, 162, 184, 215, 219

H

Hazardous Waste Management, 136, 142, 144, 147, 179, 237, 242

Hexavalent Chromium, 36, 39

Hydrogen Sulfide, 16, 170

I

Industrial Wastewater, 2, 4-5

Intermediate-level Waste, 49-50

Ion Exchange, 53

L

Land Disposal, 64, 69, 75, 242

Leachate, 14, 69-70, 116, 118, 140, 145, 170

Light Water Reactor, 53, 55

M

Manure Management, 90, 98, 111

Materials Recovery Facilities, 73, 213

Mercury, 20, 28, 30, 35-36, 38-40, 65, 73, 75, 137, 139, 182, 243

Methane, 3, 14-16, 61, 70, 77, 85, 89-91, 95, 126-127, 162, 188, 193

Mill Tailings, 48, 54

Mixed Oxide Fuel, 43-44

Municipal Food Waste, 16

Municipal Solid Waste, 1, 3, 62, 64, 66, 68-69, 71-73, 124, 179, 186, 204

N

Nuclear Power Plants, 43, 48, 172

O

Ocean Dumping, 22, 60

Organic Waste Management, 60, 121, 123

P

Pay As You Throw, 186, 225

Personal Protective Equipment, 16, 39

Pit Emptying, 80

Polyvinyl Chloride, 36, 75, 229

Precycling, 152, 162-167

R
Radioactive Waste, 1, 42-44, 46, 48, 50-58, 147-151
Reactive Oxygen Species, 36
Recycling, 2, 7-8, 13, 23-24, 26-28, 30-34, 37-40, 43, 50, 59-66, 68, 70-72, 76, 79, 89, 103, 121, 136, 144, 147, 155, 158, 162, 166, 169, 175, 179-183, 187, 194, 207, 228, 243
Reduce, Reuse, Recycle, 89, 163, 172, 175, 234
Reed Beds, 85-86
Resource Efficiency, 152, 177-178
Runoff Containment, 93, 96

S
Sanitary Landfill, 60, 69, 73
Sanitary Sewer, 4, 16-17, 126
Secure Landfill, 144-145
Septic Tank, 6, 82-84, 112, 141
Sewage Sludge, 1, 6, 9, 17-23, 68, 72, 124, 127-128, 210
Sewage Treatment, 4, 6, 17, 19-20, 86, 180
Solid Waste Management, 60, 62, 64, 72-73, 162, 169, 172, 175, 189, 230
Space Disposal, 56
Spent Nuclear Fuel, 44-46, 50, 58, 148-149
Subsurface Burial, 120-121
Surface Runoff, 4-5, 28, 69, 92, 104-105, 108, 114, 121

T
Transuranic Waste, 50, 55

U
Uranium Mill Tailings, 48

V
Vacuum Truck, 80, 83
Vermicomposting, 15, 208

W
Waste Forage, 88, 101, 116-117
Waste Hierarchy, 152, 154-155, 159, 172, 175-176
Waste Isolation Pilot Plant, 50, 54
Waste Minimization, 72, 76, 78, 142, 152, 154, 159, 186
Waste Sorting, 189-190
Waste Stabilisation Ponds, 84, 86
Water Pollution, 4, 28, 34, 90, 96, 141
Weapons-grade Plutonium, 44, 55

Z
Zero Waste Hierarchy, 175-176
Zero Waste System, 61, 175

www.ingramcontent.com/pod-product-compliance
Lightning Source LLC
LaVergne TN
LVHW080921210926
848485LV00004B/703